Henry Newell Martin

A handbook of vertebrate dissection

Henry Newell Martin

A handbook of vertebrate dissection

ISBN/EAN: 9783337257354

Printed in Europe, USA, Canada, Australia, Japan

Cover: Foto ©berggeist007 / pixelio.de

More available books at **www.hansebooks.com**

A HANDBOOK

OF

VERTEBRATE DISSECTION.

BY

H. NEWELL MARTIN, D.Sc., M.D., M.A.,
PROFESSOR IN THE JOHNS HOPKINS UNIVERSITY,

AND

WILLIAM A. MOALE, M.D.

PART I.

HOW TO DISSECT A CHELONIAN.

NEW YORK:
MACMILLAN AND CO.
1881.

COPYRIGHT, 1881,
BY
MACMILLAN & CO.

TROW'S
PRINTING AND BOOKBINDING COMPANY
201-213 *East 12th Street*
NEW YORK

PREFACE.

THE following pages, which contain directions for the practical study of the anatomy of a Chelonian, are the first instalment of a series which has had its origin in my own needs as a teacher, and which, when completed, will form a Handbook of Vertebrate Dissection.

Some years ago Professor Huxley published a little book called "Practical Biology," in the preparation of which it was my good fortune to assist him. That book was designed to give students commencing the study of biology a good knowledge, based on their own observations, of the structure and life-history of a number of typical plants and animals, with the object of laying a firm foundation for the further study of animal or vegetable morphology or physiology: it was, so far as I know, the first important educational step taken in recognition of the fact that there is a science of living beings, as such, quite apart from any division of them into animals and plants; and that the accurate observation of the phenomena presented by living forms of matter of all varieties forms the basis of one single science. This truth had, of course, been recognized long before, and was accepted by nearly all scientific naturalists, but no really important step toward organizing the teaching of biology in accordance with it had previously been taken.

In the past five years, during which I have been responsible for the direction of biological studies in the Johns Hopkins University, I have endeavored to arrange the curriculum so as to ensure that each student, before taking up any one branch of biology for special study, shall have acquired a fairly good knowledge of what may be called general biology. No undergraduate is permitted to devote his time to the study of botany, or animal morphology or physiology, until he has spent at least one year in learning something about animals and plants in general. Such a course is indisputably far better than one more specially botanical or zoölogical for those who merely desire some knowledge of biological science as a part of general education ; and I have been yearly more firmly convinced that it affords also the best beginning for the student who desires to become ultimately a botanist, zoölogist, or physiologist.

While the general structure of all plants can be fairly well understood after the careful examination of a small number of selected types, this is not the case with animals ; and I have accordingly found it necessary to include in the annual course several types, especially vertebrate, not described in the "Practical Biology." Of these a Chelonian is much the most difficult to dissect; and, partly for this reason, partly because the great persistence of vitality in its various organs after general death of the animal makes it extremely promising for physiological experiment, it was the one for whose dissection I first drew up laboratory directions for the use of my students. Finding them of great value in saving time to teacher and pupil, and also in preventing the waste of material which is apt to occur when a student sits down without specific directions to dissect an animal whose structure is almost entirely unknown to him, I thought they might be of use in other laboratories. I accordingly asked my friend and former pupil, Dr. William A. Moale, to assist me in preparing for publication detailed directions for the dissection of a number of vertebrate animals. The present pamphlet will shortly be followed by two others, containing directions for the dissection of a pigeon and a rat, both of which are well on the way toward completion. We intend ultimately to include in the series a bony and a cartilaginous fish, a lizard, and one of the large-tailed amphibia which form such a characteristic feature of the American Fauna.

In the present instalment of our work, *Pseudemys rugosa*, though far from being the most widely distributed of American Chelonia, is the species selected for description, mainly because it is sold in the markets here during the winter months, and was, therefore, readily obtainable by us. The fact that in many places this particular species may not be found is of no importance, as the end in view is not to provide a monograph on any one species, but to show a student " How to dissect a Chelonian." The members of the order are so similar in all important points that the best method of examining the structure of one species is the best also for almost any other. I am not sure, in fact, that it is not better in all cases to provide students with species slightly different from the one described ; their attention is kept more alert when they find they cannot altogether rely on the description in the book, but have to look at everything carefully for themselves.

These directions for dissection are of course not meant to be used by themselves, but to accompany lectures on the class and order of the type described, or the reading of a good text-book. Some knowledge of human osteology, which affords the best starting-point for every student of vertebrate morphology, is also assumed.

H. NEWELL MARTIN.

BALTIMORE, July 1, 1881.

CONTENTS.

―――

                                      PAGE

ZOOLOGICAL POSITION OF THE RED-BELLIED SLIDER TERRAPIN, ... 1

THE ANATOMY OF THE RED-BELLIED SLIDER TERRAPIN, . 6

INDEX, ... 89

DESCRIPTION OF THE FIGURES.

FIG. 1, represents the skull viewed from the base ; FIG. 2, as seen from the left side and partly from the front ; FIG. 3, represents the roof of the skull ; and FIG. 4, the inner side of the right half.

The references in the different figures are as follows : 1, the premaxilla; 2, the maxilla; 3, the palate bone; 4, the vomer; 5, the pterygoid ; 6, the basi-sphenoid ; 7, the basi-occipital; 8, the exoccipital; 9, the opisthotic ; 10, the squamosal ; 11, the quadrate ; 12, the quadrato-jugal ; 13, the jugal ; 14, the post-frontal ; 15, the columella auris : a thin plate of cartilage, which covered the inner division of the tympanum, has been removed to expose part of the columella ; 16, the parietal ; 18, foramen for exit of external carotid artery ; 19, the posterior nares ; 20, part of the opisthotic ; 21, occipital condyle ; 22, orbito-temporal fossa; 23, articular surface for mandible ; 25, anterior nares ; 26, supra-occipital bone ; 27, frontal bone ; 28, naso-prefrontal bone ; 29, proötic ; 30, epiotic, anchylosed to the supra-occipital ; 31, canal for entry of carotid artery ; 32, alisphenoid bone ; 33, cartilaginous inner wall of labyrinth ; 34, foramen for exit of internal carotid artery.

ZOÖLOGICAL POSITION OF THE RED-BELLIED SLIDER TERRAPIN.

SUBKINGDOM, **Vertebrata.** DIVISION, **Sauropsida.** CLASS, **Reptilia.** ORDER, **Chelonia.** FAMILY, **Emydoidæ.** GENUS, **Pseudemys.** SPECIES, **rugosa.**

Characters of the Sauropsida.*

1. Almost always an epidermic skeleton, in the form of scales or feathers.

2. The vertebral centra are ossified, but have no terminal epiphyses.

3. The skull has a completely ossified occipital segment and a large basi-sphenoid. There is no separate para-sphenoid in the adult. The proötic is always ossified, and either remains distinct from the opisthotic and epiotic during life, or unites with them only after they have anchylosed with adjacent bones.

4. There is always a single convex occipital condyle, into which the ossified ex-occipitals and basi-occipital enter in various proportions.

5. A mandible is always present, and each ramus consists of an articular ossification and several membrane-bones. The articular is connected with the rest of the skull by an ossified quadrate.

6. The apparent ankle-joint lies between the proximal and distal divisions of the tarsus.

---
* The characters given are taken, with slight modification, from Huxley.

7. The alimentary canal terminates in a cloaca.

8. The heart is trilocular or quadrilocular; some of the blood-corpuscles are red, oval, and nucleated.

9. The aortic arches may be two or more; when only one persists in the adult it is on the right side.

10. Respiration is never performed by branchiæ; and after birth always by lungs, in which the bronchi do not branch dichotomously.

11. A thoracic diaphragm may exist, but never forms a complete partition between the thoracic and abdominal viscera.

12. The cerebral hemispheres are never united by a corpus callosum.

13. The reproductive organs open into the cloaca. The oviduct is a Fallopian tube with a posterior uterine dilatation.

14. There are no mammary glands.

15. All are oviparous or ovo-viviparous.

16. The Wolffian bodies are replaced functionally by permanent kidneys.

17. The embryo has an amnion and a large respiratory allantois; and develops at the expense of a large mass of food stored up in the egg.

Characters of the Reptilia, as distinguished from **Aves,** the remaining class included in the division Sauropsida.

1. Exoskeleton composed of horny plates (epidermic), or bony plates (dermic), or both. The epidermic scales never develop into feathers.

2. The vertebral centra may be amphi- or pro- or

opistho-cœlous, or have flat faces; but their ends are never cylindroidal.

3. When a sacrum is present its vertebræ have large expanded ribs, with the ends of which the ilia articulate.

4. The sternum, when present, is rhomboidal; and, when many ribs are connected with it, the hindermost are attached to a single or double backward median prolongation. The sternum may be converted into cartilage-bone, but is never replaced by membrane-bone, and does not ossify from two or more definite centres.

5. When an interclavicle exists it remains distinct from the clavicles.

6. The manus contains more than three digits; and, at fewest, the three radial digits have claws.

7. The ilia are prolonged farther behind the acetabula than in front of them, and the inner wall of the acetabulum is wholly or almost wholly ossified. The pubes are directed ventrally and forward, and, like the ischia, meet in a ventral symphysis.

8. The digits of the pes are not fewer than three; the metatarsals are not anchylosed together or with the distal tarsal bones.

9. Not fewer than two aortic arches persist. Two arterial trunks are given off from the right ventricle, or the part of the single ventricle that answers to it. The venous and arterial currents are connected in the heart, or at the origins of the aortic arches.

10. The blood varies in temperature with the surrounding medium. There are usually two semilunar valves at the origin of the aortic and pulmonary trunks.

11. The corpora bigemina lie on the dorsal aspect of the brain. The spinal cord presents no sinus rhomboidalis.

12. The caudal vertebræ often form a series equal in length to the rest of the body.

13. Teeth are frequently present.

Characters of the Chelonia.

1. None of the vertebræ have transverse processes, and the dorsal vertebræ are not movable on one another; nor are the ribs movable upon the vertebræ. Most of the dorsal vertebræ and ribs are restrained from motion by the union of superficial bony plates, into which they pass, to form a carapace.

2. All but the soft tortoises (*Trionyx*) have part of the epidermis modified to form horny plates, which constitute the tortoise-shell. These plates are arranged in definite series on the carapace and plastron.

3. The presacral vertebræ are few.

4. No sternal ribs, and no trace of a true sternum. The ventral surface is protected by a number (usually nine) of dermal membrane bones, forming the plastron.

5. The sacrum consists of only two vertebræ; the sacral ribs are not anchylosed to their vertebræ.

6. The anterior caudal vertebræ have no transverse processes, but some have ribs.

7. In the skull, all the bones but the mandible and hyoidean arch are immovably united together: the epiotic anchyloses to the supra-occipital; the proötic and opisthotic remain as distinct bones. The alisphenoid is unossified or small, a downward prolongation of the parietal taking its place. There is no bony presphenoid or orbito-sphenoid. The mesethmoid cartilage largely persists. The orbito-temporal fossa is covered by a bony arch. The dentary pieces of the mandible are anchylosed together.

8. The pectoral arch is ossified in such a way that scapula and precoracoid form one bone, but the coracoid remains distinct.

9. The heart has a single ventricle.

10. There are no teeth, but the jaws are provided with horny cutting-plates.

THE ANATOMY OF THE RED-BELLIED SLIDER TERRAPIN (*Pseudemys rugosa*).*

1. General External Appearance.—Note the hard case in which most of the animal is enclosed, and which varies in tint considerably in different specimens; on the whole it is dark brown above, and orange-red and yellow beneath.

The case consists of a convex dorsal plate, the carapace, and a ventral flat plate, the plastron, united at the sides by bony bridges, but separate anteriorly and posteriorly, leaving large open spaces, in which can be seen certain soft parts covered by wrinkled, scaly skin, and protected by the projecting edges of the carapace and plastron.

Most of the skin is dark, with bright yellow stripes; but in the hollows which receive the hind limbs when withdrawn, and often also in the corresponding hollows for the fore limbs, it is almost white.

2. Insert a tenaculum into the soft parts beneath the head and draw out the latter; force open the mouth and direct a pipette, containing about a teaspoonful of chloroform, into the opening of the glottis, which will be found in an elevation at the base of the tongue; then blow the chloroform into the lungs. This will very soon kill the animal. Meanwhile, note the long,

---

* *Ptychemys rugosa* (Ag.). *Emys rubriventris* (Holbr.).

flexible neck, the folds of skin about the shoulders, and, after the tenaculum is removed, the way the head is withdrawn between these folds, so that only the tip of the snout is visible. So soon as the animal is dead and flaccid, you can readily stretch out the neck and proceed with your examination.

3. External Characters of Head and Neck.

a. Observe the form of the head: flattened from above down; wide posteriorly, where it is very distinctly differentiated from the neck; and narrowing toward the snout.

b. On the ventral aspect of the head and neck the *hyoidean apparatus* can be felt through the skin of the throat; and farther back the windpipe.

c. The very large *mouth opening*, provided with notched horny plates instead of teeth along the margins of the jaws; note the marked projection on each side of the median notch of the upper jaw.

d. On the extremity of the snout the openings of the *anterior nares;* a pair of round apertures placed close together.

e. The *eyes*, around which can be felt the anterior margins of the bony *orbits;* the thick, opaque upper *eyelid*, and the larger, more movable lower one; the *nictitating membrane*, or third eyelid, may be found covering the eyeball or withdrawn to the anterior corner of the eye opening; in the latter case seize its margin with forceps and draw it back over the eye so as to see it well; it is much more movable and less opaque than the eyelids proper.

f. Just behind the angle of the gape, the *tympanic membrane* will be seen as a circular slightly depressed area of skin, around the margin of which a hard supporting ring can be felt.

4. The Interior of the Mouth Cavity.

a. Open the mouth and observe that the horny plates covering the jaws extend in for some way over both floor and roof of the buccal cavity; also that each has a second notched ridge within the one forming its outer boundary. Into the groove between the two upper ridges the external lower ridge fits, and similarly the inner ridge of the upper jaw fits between the two ridges of the lower.

b. On the roof of the mouth, note that the horny plate is succeeded by diverging folds of mucous membrane, between which are seen the openings of the *posterior nares* with the back edge of the *septum narium* between them. Note also that the eyeballs do not cause projections on the roof of the mouth, as in the frog, but are completely separated from it by a hard partition.

c. On the floor of the mouth is seen the short, conical, pointed, and but slightly movable *tongue*. At the base of the tongue is a considerable elevation, in which is the longitudinal slit of the *aditus laryngis*, unprotected by an epiglottis.

EXTERNAL CHARACTERS.

d. On each side of the mouth cavity, behind the elevation caused by the mass of muscles passing down to the lower jaw, is the small opening of a *Eustachian tube*. It will be found more easily later (177, *b*).

5. The Limbs.
a. Note the manner in which the limbs, when not in use, are packed away between the projecting edges of carapace and plastron, in hollows at the sides of the neck and pelvis.
b. The *fore limbs* are flattened, clumsy, and paddle-like. When extended they curve back over the front edges of the bridges which unite carapace and plastron. The division into *brachium*, *antebrachium*, and *manus* is not obvious to the eye, but may readily be felt. The limb ends in five digits, connected by a web, and each armed with a large claw.
c. The *hind limbs* show more plainly the division into *femur*, *crus*, and *pes*. Each terminates in five digits united by a web, and possessing claws, except the fifth.
d. Seizing the pelvic region between finger and thumb, note that it has a considerable range of lateral sway. This movement takes place at the articulation between the last dorsal and the first sacral vertebra, and to some extent compensates for the rigidity of the whole dorsal region of the animal.

6. The tail, short, conical, and pointed; usually bent to one side, so as to be protected by the overlapping edge of the carapace.

On its under surface is the *cloacal opening;* a transverse slit, with puckered edges.

7. The Epidermic Exoskeleton mainly consists of horny plates (tortoise-shell), of definite number and very constant arrangement, found covering carapace and plastron.

    a. On the plastron are sixteen pieces, twelve arranged in pairs on the sides of the middle line, and the remaining four covering the margins of the bony bridges uniting carapace and plastron.

    b. On the carapace, in the middle line in front, is a small *nuchal plate;* and, meeting in the middle line behind, over the base of the tail, are a pair of *pygal plates.* Running along the middle of the back, from nuchal to pygal plates, are five unpaired *central plates.* On each side of the latter are four *centro-lateral plates;* and, external to these, eleven *marginal plates* on each side, extending along the edge of the carapace from the nuchal to the pygal plates, and, like them, curving around the edge so as to cover in the projecting margin of the carapace on its ventral side.

    c. The claws, the horny coverings of the jaws, and the scales (which are especially developed on the paddles and the dorsal side of the tail) are also parts of the epidermic exoskeleton.

8. Make drawings showing the arrangement and form of the epidermic plates on carapace and plastron. Then immerse the animal in boiling water for about two min-

THE CARAPACE.

utes, giving first another dose of chloroform (2) should there be the least sign of sensibility remaining; the tortoise-shell can then be peeled off. When it is removed, note that though the bones beneath present surface-markings corresponding to the edges of the epidermic plates, their actual number and arrangement are quite different.

9. **The Bony and Cartilaginous Skeleton.**—For the satisfactory study of the skeleton, the student should have beside him a prepared articulated skeleton for reference, and should disarticulate and examine another skeleton for himself. For this purpose boil the animal thoroughly; the plastron can usually then be separated from the carapace by seizing each and pulling them forcibly apart; if this prove impracticable, divide the uniting briḋges by a saw, then cut through the soft parts attached to the dorsal aspect of the plastron and remove the latter; clean all the bones carefully, but for the present do not separate from one another the pieces of the plastron, carapace, skull, manus, and pes. In the cleaning take great care not to injure the cartilaginous portions of the skeleton.

10. **The Carapace** is composed partly of dermic exoskeletal bony plates, and partly of true endoskeletal elements. The dermic bones are printed below in small capitals.

    a. On the exterior of the carapace, note a median row of bones succeeding one another, from before back, in the following order: the NUCHAL PLATE, rhomboidal in shape; eight *neural plates*, formed by the expanded spines of the dorsal vertebræ from the second to the ninth, inclusive; the PYGAL PLATE.

    b. On each side of the above middle row are eight *costal plates*, formed by expanded ribs attached to the dorsal vertebræ from the second to the ninth. In all except old full-grown specimens the outer ends of the costal plates are not fully ossified; and, accordingly, spaces filled only by cartilage are then left between each one and its successor.

    c. The MARGINAL PLATES, eleven on each side, extend from the nuchal to the pygal. From the second to the fifth, inclusive, they reach ventrally to articulate with the plastron.

11. On the inner (ventral) aspect of the carapace, note the projecting *centra* of **the Dorsal Vertebræ.** The most anterior is attached dorsally to the nuchal plate; the eight succeeding ones bear, dorsally, the expanded neural spines already seen on the exterior of the carapace as the neural plates.

    a. The first dorsal vertebra is comparatively short and thick, and bears anteriorly a pair of large stout *prezygapophyses* (*anterior articular processes*), which slope downward and forward.

    b. The centra of the succeeding dorsal vertebræ, back to the eighth, are considerably elongated in the antero-posterior axis, and compressed from side to side. Each has flattened ends, and between each pair lies a thick plate of cartilage, except in quite old specimens in which the intervening cartilage is much thinner. The whole set of dorsal vertebral centra, with these

THE CARAPACE. 13

plates, thus forms a continuous osseo-cartilaginous rod.

12. Examine the Ribs.—Attached opposite the interspaces between the vertebral bodies from the first to the fourth dorsal, and articulating either only with the interposed cartilage, or also with the posterior end of one centrum and the anterior end of the next, are the heads (*capitula*) of the second, third, and fourth ribs.

   a. From the capitula, thin bony bars, which may conveniently be called the *necks* of the ribs, run outward and dorsally for a short distance, and then fuse with the ventral aspect of the *costal plates* already mentioned (10, *b*).

   b. The costal plates by their inner ends articulate with the neural plates of the two vertebræ between which the head of the corresponding rib lies.

13. The arrangement of the fifth, sixth, seventh, and eighth ribs is like that described above (12), except that each articulates with the centrum of only one dorsal vertebra (that which corresponds in number with the rib); and the costal plate of the ninth rib articulates only with the neural plate of its own vertebra. In old specimens, also, the sides of the centra of the posterior dorsal vertebræ bear distinct *capitular processes*, to the ends of which the heads of the ribs are attached.

14. The ribs of the first pair are very small; they run backward and outward from the sides of the centrum of the first dorsal vertebra, and fuse at their outer ends with the costal plates of the second dorsal vertebra, the first vertebra having no costal plates of its own.

15. The tenth dorsal vertebra is comparatively wide and short, and carries a pair of small ribs which have no costal plates of their own, but anchylose with the eighth pair of costal plates (*i.e.*, those of the ninth dorsal vertebra). On the ventral aspect of each of these last costal plates is to be seen a small elevation, over which the sacral ribs (16) glide in lateral movements of the pelvis.

16. The two **Sacral Vertebræ** may be found still attached to the carapace, or one or both may have come away with the tail in preparing the specimen. Each has a pair of short ribs, which run outward and end in cartilages—which, in the articulated specimen, will be seen to articulate with the iliac bones (71, *a*). That end of each rib next the vertebral centrum is also cartilaginous. The anterior rib is much stouter than the other.

17. Take the carapace to bits, piece by piece, noting meanwhile the finely dentated *sutures* by which most of the bones articulate. As you separate the pieces, note also that—

    a. The neural arch of the first dorsal vertebra is much shorter than the centrum, which is overlapped in front by the posterior end of the nuchal plate, and behind by the neural plate of the succeeding vertebra, the spine of the first dorsal vertebra not expanding to make any neural plate. This vertebra also has not a persistent *neuro-central suture*, the neural arch being firmly grown into continuity with the centrum. Its remarkable prezygapophyses, which articulate with the *postzygapophyses* (*posterior articular processes*) of the last cervical vertebra, have already been mentioned (11, *a*).

b. The centrum of the second dorsal vertebra is covered on the dorsal side, for its anterior two-thirds, by its own neural arch, the two uniting by a persistent *neuro-central suture;* the posterior third of the centrum is overlapped by, and articulates with, the neural arch of the vertebra behind. Its neural spine expands to form the first neural plate of the carapace.

c. The arrangement of the parts of the dorsal vertebræ from the third to the eighth, inclusive, is pretty much like that found in the second; but in each the neural plate reaches as far back as the posterior margin of the centrum, although the *laminæ* of the neural arch of the ninth vertebra overlap the posterior part of the centrum of the eighth.

d. The neural plate of the ninth dorsal vertebra reaches back and overlaps the tenth vertebra, which has no neural plate of its own.

18. Examine carefully a **Costal Plate**—the third, for example. Note the dentated sutures on each of its margins; the continuity of the neck of a rib with its ventral surface; the process which extends from its outer end to be inserted into a pit in the corresponding marginal plate. In partly grown specimens this process is long and conspicuous, the portions of costal plate in front of and posterior to it in that region being not yet ossified (10, *b*).

Also, examine more carefully the marginal plates now that they are separated ; and note their relations to the various costal plates.

19. The Plastron is composed entirely of dermic bones. It consists of nine pieces : one median unpaired bone toward its anterior end, called the ENTOPLASTRON ; and eight bones arranged in pairs on each side of the median line ; those on each side, from before back, are named successively, EPIPLASTRON (? clavicle), HYOPLASTRON, HYPOPLASTRON, and XIPHIPLASTRON. The outward ends of the hyo- and hypo-plastron curve dorsally, and make up most of the bony bridges uniting carapace and plastron.

20. The Cervical Vertebræ are eight in number and freely movable, except the first and second. In general, each consists of a centrum and neural arch, with pre- and post-zygapophyses ; and the articular surfaces of the prezygapophyses look dorsally, and those of the postzygapophyses ventrally ; there are no transverse processes, but the arches present more or less of a median dorsal ridge, representing a rudimentary spinous process. The posterior end of the body is produced backward and ventrally some distance beyond the neural arch ; and the intervertebral foramina are arched over by the articular processes. The centra bear a slight median ridge on the ventral aspect, and are thick and cylindrical near the anterior end, but flattened dorso-ventrally at the posterior.

THE CERVICAL VERTEBRÆ. 17

21. In individual cervical vertebræ, note the following peculiarities :
  a. The *first* (*atlas*) is a ring-like mass, composed of three pieces : a ventral and two dorso-lateral, united by cartilages. Its anterior end presents, where the three pieces meet, a deep pit, in which the occipital condyle (25, *a*) plays ; posteriorly is a similar pit for the odontoid process of the axis (21, *b*).
  b. The *second cervical vertebra* (*axis*) has a large *odontoid process*, attached by cartilage to the anterior end of its centrum.
  c. The *postzygapophyses* of the *seventh* and *eighth cervical vertebræ* are remarkably large and stout, and those of the eighth arch backward to articulate with the prezygapophyses of the first dorsal vertebra (11, *a*). These two vertebræ are also peculiar in their neural spines, which consist of a pair of ridges which meet anteriorly, but diverge behind to run out over the postzygapophyses.
  d. The variety in the surfaces by which the centra of the cervical vertebræ articulate with one another. The second and third vertebræ are *opistho-cœlous*; the fourth has a convexity on each end of its centrum ; the fifth is concave anteriorly, and has a pair of convexities behind. The sixth has two concavities in front, and two convexities behind. The seventh has two concavities in front, and two behind. The eighth has two convexities in front and two behind.

22. The Caudal Vertebræ are freely movable, and in the anterior part of the tail procœlous. From the fifth to the tenth, inclusive, they have well-marked rudimentary ribs attached to them. None have transverse processes. Becoming smaller and smaller toward the tip of the tail, they are finally reduced to minute subcylindrical masses, without any processes, but each still possessing a centrum and a neural arch enclosing a neural canal.

23. The Skull.—After boiling the skull, clean away the soft parts, taking care to leave, in so doing, as much of its cartilaginous parts as possible—the thin plate between the nostrils is especially apt to be removed. When cleaned, make outline drawings of the skull, as seen from behind, from the base, and antero-laterally.

24. The skull is flattened from above down, and wide from side to side. Note that the brain-case proper is much narrower than the skull, which is expanded posteriorly, to form a pair of large *post-auditory processes*, from which two bony arches proceed forward on each side : one below the orbit to the upper jaw, and the other behind the orbit to the brain-case proper. The *temporal fossa* is covered in by this arch on each side, and is continuous with the *orbital cavity*, so that the two form one *orbito-temporal fossa*.

25. On the Posterior Aspect of the Skull, note :
    a.   The *foramen magnum*, and below it the single *occipital condyle*.
    b.   On each side of the condyle are two small foramina ; through the more mesial of these, the *hypoglossal nerve* (180, *i*) passes out.

THE SKULL.

c. External to these is the larger *foramen lacerum posterius*, which gives exit to the pneumogastric and spinal accessory nerves (180, *h*).
d. Still farther out is seen the conical *post-auditory process*, covered over by the *squamosal bone*.
e. Above the foramen magnum the bony spine of the *supra-occipital bone* projects backward; on each side of this the roof of the skull is concave, and in the bottom of the hollow is a large foramen, through which the *external carotid artery* passes out to supply the exterior of the head.

26. On the Base of the Skull, note:
a. The ventrally projecting *quadrate bone*, with its terminal facet for articulation with the mandible.
b. Farther forward, and near the middle line the *posterior nares*, and somewhat behind and to the outside of these, the aperture on each side for the transmission of a branch of the fifth cranial nerve.
c. In front of the quadrate a large aperture leads up into the orbito-temporal fossa; through it the main muscles of mastication formerly passed to the lower jaw from their origins on the sides of the brain-case.
d. Behind the quadrate is a plate of cartilage, ossified in some specimens, which is often removed in cleaning the skull; if this has not happened remove it now on one side. This will lay open part of the labyrinth and the inner section of the tympanic cavity, which will be

seen to communicate with an outer division by means of a narrow passage, through which the *columella auris* (now exposed *in situ*) passes from the tympanic membrane on its way to the *fenestra ovalis*.

e. Just outside the cartilaginous plate referred to above and close behind the quadrate is a foramen for the entry of the carotid artery, which, within the bone, divides into an *internal carotid*, which enters the skull cavity, and an *external carotid*, which emerges by the foramen on the roof of the skull mentioned above (25, *e*).

f. In front of the cartilage, and nearer its outer than its inner side, is a small aperture which leads into a canal which passes forward some way in the bone forming the base of the skull. Into it one division of the anterior extremity of the sympathetic nerve will subsequently (199) be found to pass.

g. External to the aperture of the sympathetic channel, and in front of the aperture of the carotid canal, is the small foramen of exit for the seventh cranial nerve (200).

27. **On the Side of the Skull,** note :

a. The large ear-opening leading into the external division of the tympanum, which reaches back into the post-auditory process. Close to the outer side of the proximal end of the quadrate bone is the narrow passage through which the columella passes on to the inner expansion of the tympanic cavity.

b. Farther forward is the *temporal fossa* covered

THE SKULL. 21

over by the arches already mentioned (24). In it, just above the inner side of the proximal end of the quadrate bone, is the large foramen through which the *superior* and *inferior maxillary divisions* of the fifth cranial nerve leave the skull (190, 191).

28. Note that the anterior end of the brain-case proper is not inclosed by bone, but opens widely between the orbits ; also, that between the orbits there is no bony partition, the *inter-orbital septum* being nearly entirely membranous, and usually removed in cleaning the skull. In the middle line between the orbits and close to the bone bounding them below, will, however, be found the *naso-ethmoidal cartilage*, passing out from the interior of the skull-chamber and there flattened dorso-ventrally ; between the orbits it becomes cylindrical, and then, farther forward, expands into a vertical plate (*mesethmoid*), which enters the nose and forms most of the partition between the nostrils.

29. The floor of the orbit is not completely bony, but presents a considerable *fontanelle*, closed in the recent state by membrane : through this a branch of the superior maxillary nerve will subsequently (190) be found to pass.

30. On the posterior boundary of the nares, note a round aperture above the mesethmoid cartilage, and close to the roof of the nose. Into it, in the recent state, reached an anterior prolongation of the brain, enveloped in *dura mater* (169), which passed out of the large anterior opening of the proper brain-case above noted (28), and, thence, forward between the eyeballs and close to the bones on the roof of this part of the skull.

31. If necessary, now boil the skull again until its bones are readily separable. Then, working with the articulated specimen beside you for reference, remove the bones, one by one, in the order in which they are described below; as each is removed, and its boundaries thoroughly ascertained, sketch it in on your outline drawings (23).

32. The *post-frontal* is an irregular, squarish bone which forms most of the posterior margin of the orbit; it articulates with the *frontal* and *parietal* above, and *jugal* and *quadrato-jugal* below. A long process reaches back from its postero-inferior margin.

33. Below the post-frontal and forming the posterior margin of the orbit, is the *jugal*, squarish as seen from the exterior, but sending inward a process which forms part of the floor of the orbit and reaches internally to the pterygoid. It is overlapped by the maxilla in front. The jugal overlaps behind the bone next described.

34. The *quadrato-jugal* bounds the anterior margin of the tympanic cavity. It sends downward and backward a slender process, which turns around the margin of the ear-opening for a variable extent. Posteriorly it joins the quadrate below and the squamosal above.

35. Behind the tympanum is a conical cap of bone, the *squamosal*, which covers a similarly shaped semi-

cartilaginous cap, whose hollow is continuous with the tympanic cavity. In front, the squamosal articulates with the quadrato-jugal and the quadrate ; and, behind, it overlaps and conceals parts of the quadrate and opisthotic bones.

36. The *quadrate* forms a large, irregular, bony mass, in which lies much of the tympanic cavity. It is produced ventrally to articulate with the mandible. It articulates at its upper end, on the inner side, with the proötic and opisthotic bones. In removing the quadrate, take care not to displace the columella auris.

37. The *opisthotic*, which protects the posterior part of the inner ear, may now be removed and examined ; it forms the posterior boundary of the fenestra ovalis (26, *d*). Behind, it articulates with the exoccipital ; above, with the supra-occipital ; and in front with the proötic. It also articulates with the quadrate, which conceals a great portion of it in the natural positions of the parts.

38. In front of the opisthotic is the *proötic*, bounding the fenestra ovalis in front. It articulates with the last bone ; also in the base of the skull with the basisphenoid and pterygoid ; and above with the parietal and epiotic.

39. The *epiotic* lies above the proötic and opisthotic, with both of which it articulates. In adults it is not separable from the supra-occipital.

40. The *supra-occipital bone* bounds the foramen magnum above ; in front it is overlapped by the parietals ;

below it articulates with the exoccipitals and anchyloses with the epiotic.

41. The *exoccipital bones* form the lateral boundaries of the foramen magnum, and, together with the basi-occipital, form the occipital condyle.

42. The *basi-occipital* articulates on its sides with the exoccipitals, and anteriorly with the basi-sphenoid.

43. On the base of the skull, in front of the basi-occipital, is seen the *basi-sphenoid*. It articulates with the pterygoids in front, and the proötics on its sides; when removed, observe that its upper or inner surface presents anteriorly a concavity (*sella turcica*) bounded by ridges, which rise up in front as a pair of thin bony plates, the *anterior clinoid processes*. Posteriorly its upper surface presents a median ridge, continuous with a similar one on the corresponding surface of the basi-occipital.

44. In front of the basi-sphenoid, and partly covering its ventral surface, are the *pterygoid bones*.

45. The *parietal bones* form a large part of the roof of the cranium. Posteriorly they overlap the supra-occipitals; in front they unite with the frontals; in the middle line they meet one another; laterally, they reach down and cover in much of the sides of the brain-case proper, meeting the pterygoids, palate bones, and proötics.

46. The great downward extension of the parietals is a marked feature of the Chelonian skull, causing them to cover in a region which, in most cases, is protected by the *alisphenoid*. This bone is small and closely attached to the middle third of the ventral margin of the parietal in the skull of the Slider.

47. The *palate bones* lie in front of the pterygoids. They partly bound the posterior nares above, and also form a small part of the roof of the mouth on the sides of the recess in which the posterior nares open. Above they meet the pterygoids, and externally the superior maxilla. Between them lies—

48. *The vomer*, a slender bone having on its anterior extremity a knob which appears on the roof of the mouth. With the palate bones it completes the roof of the posterior nares.

49. Meeting in the middle line at the front of the skull are the small *premaxillæ*, which bear part of the horny cutting-ridges of the upper jaw.

50. The *maxillæ* lie on the outer sides of the premaxillæ, and form the greater part of the roof of the mouth and of the upper boundary of the gape; they also form a large part of the floors of the orbits.

51. The *naso-prefrontal bones* roof in the front part of the nose chamber. In front they join the maxillæ, and help to complete the orbital boundary. Behind they join—

52. The *frontal bones*, which meet in the middle line above the orbits. Behind, these articulate with the parietal and post-frontal bones, and in front with the naso-prefrontal bones.

53. The Mandible is a V-shaped mass, composed of several bones, and presents an anterior dorso-ventrally compressed *body*, and two lateral *rami*, one at the posterior end of each half of the body, and compressed laterally. On the upper border, on each side, is a prominent *coronoid process* where ramus and body meet; and on the free end of the ramus is an oblique surface, with a central elevation, which articulates with the ventral end of the quadrate (36), which has a corresponding groove into which the ridge fits, so that a firm hinge-joint is formed between them. On the upper surface of the body are ridges, which bear the horny cutting plates of the lower jaw (4, *a*). On the upper edge of the ramus is a fissure leading into the mandible; a branch of the inferior maxillary nerve (191) enters there. On the inner side of the ramus are two foramina; through the lower a branch of the above nerve passes out into the floor of the mouth.

54. The following parts enter into the composition of the mandible; to separate them needs some patience, as the bones adhere very tightly and the lines of articulation are not very visible.

    a. On the posterior aspect of the bony body of the mandible is a groove, filled in the recent skeleton by a cartilaginous rod (*Meckel's cartilage*), which will subsequently be found to extend back through the rami.

    b. Take hold of the articular end of one ramus, and work it to and fro until it is loose. Then carefully detach it, disturbing the remaining bones as little as possible. It will bring with it Meckel's cartilage, which will be found to join its fellow in the middle line in front; divide the cartilage at this point, and remove the whole loosened piece; it consists of four bones, which are to be separated and examined.

c. One of these pieces, placed to the outside, and forming the outer boundary of the fissure above mentioned (53), is a thin bone, lying in the vertical plane; it is the *surangular*. In the natural position of the parts it is largely overlapped by the dentary (54, *h*).

d. Forming the inner boundary of the fissure, and reaching back to form part of the surface which articulates with the quadrate, is another thin, flat bone, placed vertically: it is the *splenial*, and forms nearly all the inner surface of the ramus of the mandible.

e. Wedged in between the lower edges of the splenial and surangular is a narrow bony bar, the *os angulare*.

f. When the above pieces are removed, Meckel's cartilage will be left, with a small polygonal bone continuous with its posterior end. This bone is the *os articulare*. It forms most of the articular facet of the mandible, and is the only part of it ossified in cartilage.

g. The part of the mandible left when the above loosened portion (54, *b*) was removed, contains two bones: one is the *coronoid*, which forms the process of that name (53), and articulates in front with—

h. The *dentary*, which is anchylosed to its fellow in the middle line in front, and also sends back a process on the outside of the ramus, overlapping the surangular.

55. The bones of the skull, seen in their natural relative positions from the interior of the cranial cavity, may be studied either on one-half of a prepared median dorso-ventral section of the cranium, or, better, on a recent specimen prepared by the student; as the cartilaginous parts, which are important, are apt to be removed unless special care has been taken in preparing the specimen.

56. On the Interior of the Skull, as viewed from one side, note:

- *a.* The *basi-occipital bone*, divided by the saw in the median line behind.
- *b.* External to the basi-occipital, the *exoccipital*, having in it the foramen (25, *b*) for the twelfth cranial nerve; between it and the basi-occipital another foramen; and between it and the opisthotic the foramen of exit for the vagus and spinal accessory nerves (25, *c*).
- *c.* Above the exoccipital the large *supra-occipital*, with its spine.
- *d.* On the base of the skull, and in front of the basi-occipital, the divided *basi-sphenoid*, with its median ridge and one of the clinoid processes described above (43); and in front of the clinoid process a hollow from which some muscles of the eyeball (184) arise.
- *e.* External and dorsal to the outer edge of the posterior portion of the basi-sphenoid is a cartilaginous plate (already removed if care has

THE SKULL. 29

not been taken in preparing the skull), which separates the cranial cavity from the internal ear. The extent of this plate varies very much, being less in older specimens. The corner of the supra-occipital which bounds it above, represents the *epiotic bone* (39).

f. Behind the cartilage is the *opisthotic bone*, and in front of it the *proötic*.

g. In front of the supra-occipital and proötic, and also reaching back so as to cover in a part of the supra-occipital, is the *parietal bone*, and between it and the proötic, the large foramen (*foramen ovale*) for the exit of the main divisions of the trigeminal nerve.

h. External to the anterior clinoid process of the basi-sphenoid, and in the proötic bone, is seen the opening of the canal through which the internal carotid artery enters the cranial cavity. Pass a probe back through the canal to the point where the common carotid artery enters the skull cavity (26, *e*).

i. In front of the basi-sphenoid is seen the *pterygoid bone*, forming that part of the base of the skull.

k. External to the pterygoid, below the parietal, and in front of the proötic, is the small *alisphenoid* bone.

l. In front of the descending plate of the parietal (45) is the large aperture left by the removal of the membranous interorbital septum in preparing the skull. Bounding it behind is the parietal; above, the frontal and the naso-prefrontal bones; in front, the naso-prefrontal; below, the palate bone and the vomer, and, lying on these, the naso-ethmoid cartilaginous rod (28).

m. In front of the naso-prefrontal bone is seen the nostril cavity, bounded below by the maxilla and the premaxilla ; externally by the maxilla and the naso-prefrontal bone ; and above, by the latter bone.

57. The Hyoidean Apparatus is largely cartilaginous, except in old specimens. It presents—

a. A median piece, or *body,* with a concavity on its dorsal side into which the larynx and the anterior end of the trachea fit. The ventral surface of the body of the hyoid is convex, and its most prominent median part is composed of a connective-tissue membrane. From its anterior end arises a small median pointed process, which, in the recent state, enters the tongue. To each antero-lateral angle a small cartilaginous nodule is attached by a fibrous membrane.

b. Attached to the sides of the body of the hyoid, about half way between its anterior and posterior ends, are the curved *anterior cornua,* which ossify earlier than the rest of the hyoidean apparatus. They run back, in their normal position, parallel with and close to the posterior border of the mandible.

c. The postero-lateral angles of the body of the hyoid form diverging processes to which the *posterior cornua* are attached ; they are shorter and stouter than the anterior cornua, and are curved in the opposite direction.

THE FORELIMB.

58. The Pectoral Arch consists on each side of two pieces. On the articulated skeleton note its general position with respect to carapace and plastron, and the attachment of the forelimbs to its outer angles.

59. One of the pieces of the pectoral arch on each side (*the scapulo-precoracoid*) is V-shaped, with the angle of the V turned outward. Its dorsal limb (*scapula*) is longer than the other and attached to the ventral aspect of the first costal plate, opposite the first dorsal vertebra.

60. The *precoracoid*, which forms the shorter and ventral limb of the V, nearly meets its fellow in the middle line opposite the entoplastron (19); in the recent state, stout ligaments attach it to the plastron.

61. The remaining bony piece of the pectoral arch on each side is the *coracoid bone;* it is somewhat flattened dorso-ventrally, and passes back in a nearly horizontal plane from its articulation with the scapulo-precoracoid where the two limbs of the V formed by the latter meet. Its posterior end is expanded.

62. Note the deep *glenoid fossa* where the elements of the pectoral arch meet.

63. When you open another terrapin (77) observe the stout ligament joining the distal ends of coracoid and precoracoid; it represents the *epicoracoid*, found as a distinct bone in some reptilia.

64. The Forelimb is short and stout. Note in the articulated skeleton that it lies in a nearly horizontal plane, and, in its position of rest, with the humerus directed forward and the forearm and manus backward; also that the two bones of the forearm lie in a dorso-ventral plane, the ulna being dorsal. If the forelimb were stretched out so that its axis was perpendicular to the median antero-posterior dorso-ventral plane of the trunk the thumb side would be ventral.

65. *The humerus* has a short, thick shaft with expanded articular extremities. The proximal extremity, set on to the shaft nearly at a right angle, has two very large tuberosities, and articulates with the glenoid fossa. On the distal end is the articular surface for radius and ulna.

66. *The bones of the forearm* (*radius* and *ulna*) are separate and flattened, and somewhat constricted in their median portions. The ulna is thicker and larger than the radius, and reaches beyond it at the elbow; at the wrist, the reverse is the case.

67. *The carpus* is composed of eight bones. Three lie at the distal end of the ulna; of them, that on the ulnar side, considerably smaller than the others (and often removed in preparing the skeleton), is the *pisiform;* next it is the *cuneiform* (*os ulnare*) which articulates on its proximal side with the ulna; the third bone of the proximal row is the *semilunar* (*os intermedium*), which articulates above with the ulna, and on its radial side with the side of the projecting end of the radius.

68. The distal end of the radius articulates with the proximal end of the largest bone in the carpus, which

probably represents the *scaphoid* (*os radiale*) and *os centrale* combined.

69. The four remaining bones of the carpus (*carpalia*) lie in a row at the bases of the digits; those corresponding to the first, second, and third metacarpal bones (*trapezium*, *trapezoid*, and *os magnum* respectively) are separate; the carpalia of the fourth and fifth digits, on the other hand, are united to form a single bone (*unciform*) which, however, presents a deep groove; this indicates its primitive doubleness, and will lead to the belief that it is made up of two distinct pieces, unless careful examination be made.

70. Each of the five *digits* consists of a metacarpal bone and three phalanges, except the first (*pollex*) and the fifth (*minimum*), which each have but two phalanges. The terminal phalanx on each is long, curved, pointed, and claw-like.

71. The Pelvic Girdle is composed of *the iliac*, *the ischial*, and *the pubic bone* on each side.

   a. The *ilium* is attached to the sacral ribs, as described above (16); from there the bone runs ventro-mesially and joins the ischium and os pubis at the *acetabulum*.

   b. The *pubic bones* are flat plates, which meet ventrally in a mesial *pubic symphysis*, from which a cartilaginous plate proceeds forward in the abdominal wall. External to this each bone has a large, pointed process, directed forward.

   c. The *ischial bones* meet also in a median ventral symphysis, and are much like the pubics in form, though smaller. Each reaches forward so as to meet a process sent back from the pubic bone on the same side, and enclose a large *obturator foramen.*

72. The Hind Limb, larger and somewhat stouter than the anterior, resembles it closely in general form and in its position with reference to the axial skeleton (64).

73. *The femur* is larger than the humerus, but remarkably like it in shape.

74. In the *crus* the *tibia* and *fibula* are distinct, and each enters into the knee- and ankle-joints. The tibia is larger than the fibula, and placed on its ventral side.

75. *The tarsus* has two bones in its proximal row. The larger of these (*astragalus*), placed on the tibial side, is made up of a coalesced *tibiale* and *intermedium.* The other bone in this row is the *os fibulare* (*calcaneum*). The distal row of the tarsus presents five bones (*tarsalia*), of which that on the fibular side is much larger than the others, and is regarded by some morphologists as a metatarsal bone. If this be so, the proper tarsalia are four in number: one each for the first, second, and third digits, corresponding to the internal, middle, and external cuneiform bones of human osteology; while for the fourth and fifth digits there will be a single tarsal bone, corresponding to the cuboid.

76. Each *digit of the foot* has a metatarsal bone; but, accepting the above view as to the homology of the fifth bone in the distal row of the tarsus, the fifth metatarsal is very different in form from the rest. The great toe

DISSECTION OF THE VISCERAL CAVITY. 35

(*hallux*), has two *phalanges*; the second, third, and fourth toes three each, the terminal bone on each being long and claw-like. The fifth digit has, according to the homology accepted for the bone on the fibular side of the distal row of the tarsus, either three or two phalanges; the terminal one, in correspondence with the absence of a claw on it (5, *c*), is rudimentary.

77. General Dissection of the Pleuro-peritoneal Cavity. Kill a terrapin with chloroform in the manner above described (2); saw through the bony bridges uniting carapace and plastron, and, with scissors, divide the skin in front and behind, close to the plastron, and then placing the animal on its dorsum, gradually raise the plastron from before back. Dissect away, with a scalpel, the soft parts attached to its inner surface; cut close to the bone, so as to avoid, as far as possible, all injury to the soft parts. As you separate the plastron, note the stout ligaments passing to it from the precoracoids. When the plastron is removed note the fibrous periosteum covering it on its dorsal side, and the attachments of muscles to it.

78. On the ventral aspect of the animal, as exposed by removing the plastron, observe:
    a. The semi-transparent pericardium, and, seen through it, the heart, which may still be beating.
    b. Anteriorly, the muscles on the pectoral arch, and the bones of the latter which can be felt through them; posteriorly, the pelvic arch, which can be likewise felt through the muscles over it.

c. The *great pectoral muscle* on each side; flat, and made up of two portions—an anterior, which will be seen to have arisen from the plastron opposite the pectoral arch, and a posterior larger portion, which arose from the plastron considerably farther back. Both parts converge to their insertion on one of the tuberosities of the humerus (65).

d. On the pelvis, a fan-shaped muscle, the origin of which from the plastron has been divided in removing the latter; the anterior fibres (*attrahens pelvis*) run backward, the posterior (*retrahens pelvis*) forward, to be attached to the large process on the antero-lateral margin of the pubis. This muscle is the homologue of the *pyramidalis* of human myology.

e. On either side of the pectoral muscles, and between them, portions of *peritoneum* are seen, covered here and there (as is also the ventral surface of the pericardium) with more or less yellow fat; in front of the pelvis the *anterior abdominal veins* are very conspicuous, lying on the peritoneum.

79. Dissect away carefully the great pectoral muscles and the pyramidales, noting the nerves which enter their deep surfaces. Take care to avoid the branches entering the anterior abdominal veins in front and behind, and note:

a. Through the pericardium and peritoneum, the large quantities of lymph contained in the cavities they enclose.

b. On the peritoneum, the two **Anterior Abdominal**

THE RENI-PORTAL VEIN. 37

Veins, turning dorsally a short distance behind the heart, and entering the peritoneal cavity. Joining each, just at this point, is a vein from the pectoral region, and uniting them in front of the pelvis, and also behind the heart, are cross-branches; that behind the heart will be seen to receive small pericardiac branches.

c. Posteriorly, note the formation of each anterior abdominal vein by the union of a small branch, running from the ventral aspect of the pelvis, with a larger vessel appearing at its outer side; the former results from the division of a single larger median vein, which, when traced back, is seen to be formed by the union of two lateral trunks with a median one; these all carry back some blood from the tail, and from the ventral regions of the pelvic girdle.

d. Trace back the outer branch entering into the anterior abdominal trunk; on the muscles covering the ilium it is joined by a large vein from the hind limb; this, the *femoral vein*, here divides into the trunk just described which enters the anterior abdominal, and another, the *reni-portal vein*, which turns dorsally.

80. If proper care has been taken in the previous dissection the veins will be filled with blood and easily traced; if they are empty they collapse, and are hard to follow, and their relative sizes are also impossible to appreciate. If this be so, cut a hole in one of the smaller branches; take a fine blow-pipe, and direct a current of air on the vessel till it strikes the aperture and distends it, so that it is readily seen; then put in the tip of the blow-pipe, and inflate the veins with air. By repeating this procedure from time to time, as required, the dissection of the venous system is greatly facilitated.

81. Trace the *reni-portal division* of the femoral vein on one side till it penetrates the muscles lying in the groin-hollow. Then, turning the leg and tail over to the opposite side, divide the skin between them and the carapace, and search in the hollow between leg and thigh for the large *caudal vein;* trace this up till it joins the reni-portal.

82. Beginning a short way behind the heart, divide the peritoneum in the middle line from before back ; a short distance in front of the pelvis it turns away toward the dorsal side, so that the peritoneal cavity does not extend to the posterior end of the trunk, nor does the peritoneal membrane line the pelvis, being merely reflected over the anterior end of the *urinary bladder*, which will be seen projecting forward from the pelvic cavity.

83. Remove the ventral portions of the peritoneum and pericardium, cutting the former away carefully from around the anterior abdominal veins and the branches entering them, so as to leave the vessels intact. Note the *bladder veins*, which enter the cross branch uniting the two anterior abdominal trunks near the pelvis, (79, *b*).

84. Putting your thumbs under the coracoid bones, turn both halves of the pectoral arch forward, tearing the ligamentous attachment of the scapula to the cara-

pace. If necessary, gently run some water among the viscera to replace the abundant lymph normally present and float the organs into their normal positions; if the lungs are collapsed inflate them moderately, by means of a blow-pipe inserted into the glottis; then sketch the parts as they lie.

85. The following points may be observed before disturbing the viscera:

   a. Anterior to the heart, and disappearing to its dorsal side, part of the *trachea* is seen. Note, that when the neck is retracted it makes a loop to the left, which is straightened out when the neck is extended.

   b. The heart, with its transversely elongated *ventricle*, and *right* and *left auricles;* these, together, almost conceal the *venous sinus*, which, however, projects slightly beyond the outer anterior corner of each auricle. Springing from the base of the ventricle are several large arteries.

   c. On each side of the heart the *right* and *left liver lobes*, with the *gall-bladder* appearing from beneath the posterior border of the right lobe.

   d. Partly covered by the left lobe of the liver, but seen beyond its edge, is the U-shaped *stomach*, ending in the *pylorus* just behind the heart.

   e. The *duodenum*, starting at the pyloric end of the stomach, passes to the right, and then dips away to the dorsal surface. Farther back are seen some coils of the other portions of the *intestine*, especially on the right side of the abdominal cavity.

f. On each side of the intestinal convolutions are seen the posterior ends of the lungs.

g. Partly covered by the pubis is the bilobed *urinary bladder*, generally much distended. If collapsed, inflate it by means of a blowpipe inserted through the cloaca.

h. Immediately posterior to the gall-bladder, a portion of the *pancreas* will be seen lying along the duodenum.

i. Note the entry of each anterior abdominal vein into the liver-lobe on its own side.

86. Next raise the right liver-lobe, and after observing the gall-bladder more closely, push the heart up, and gently separate the left liver-lobe from the stomach ; the great transverse **Portal Vein** will then be exposed, running across the liver from side to side ; opposite the heart it gets a large vein from the pancreas, and where it lies against the stomach many small veins enter it from that organ. Note, also, the bridge of hepatic tissue uniting the two main liver-lobes across the middle line, behind the heart ; and the deep fissure on the left lobe into which the right side of the stomach fits, and where many small gastric veins directly enter the liver.

87. Without cutting the mesentery, gently uncoil the **Small Intestine**, and trace it until it joins the **Large Intestine**; a small **Cœcum** projects beyond the point of communication. Raising the gut, note the **Mesentery** which slings it, and in this the numerous arches formed by branches of the mesenteric artery, from which smaller twigs pass to the intestine ; also the accompanying mesenteric veins.

THE HEART. 41

88. In connection with the first coil of the intestine, note the elongated yellowish **Pancreas**, doubled on itself; and, somewhat farther back, lying in the mesentery near where the small and large intestines join, the dark-red, soft, rounded **Spleen**. Near this organ, the breaking up of the mesenteric artery to form its numerous primary branches will be found, and the union of the mesenteric veins to form the main radicle of the portal trunk.

89. Trace the **Cystic Duct**, which may readily be seen on squeezing some of the green bile into it from the bladder, from the gall-bladder to the duodenum. It opens into the duodenum obliquely after running close along it for some distance. The **Pancreatic Duct** opens into the duodenum about an inch nearer the stomach than the bile-duct; it may easily be found on gently separating the pancreas from the intestine.

90. Trace the **Large Intestine** to the pelvis; note the dilated **Rectum**, which forms its terminal portion.

91. Examine **the Heart** more carefully. Its two *auricles* about equal in size, and the larger *ventricle*, usually paralyzed by the chloroform and greatly dilated. Divide the ligament passing from the posterior border of the ventricle to the pericardium, and raise the former so as to see the transversely elongated *venous sinus* behind it.

92. Continuous with the right end of the venous sinus, is the **right superior cava**. Trace it for about an inch from the heart, where a small vein will be seen to enter its posterior side; make an incision in this and

inflate with a blow-pipe, so as to distend the heart and vessels connected with it. When the organ is blown up the relative size and the form of its parts can be much better seen than was possible previously.

93. Follow now the **superior cava** on each side, shading insensibly into the venous sinus, and formed a short way from the heart by the union of three large trunks (118). Entering the posterior border of the venous sinus, toward the left, is the **left hepatic vein**; at the corresponding point on the right side opens a much larger trunk, formed by the junction within the right liver-lobe of the *inferior cava* and the **right hepatic vein**.

94. Clear away carefully the *lymph sinuses* and loose areolar tissue surrounding the great vessels springing from the base of the heart, and carefully clean and separate these vessels down to their origin. Extend the neck and remove the skin and connective tissue covering its ventral surface and sides.

Three vessels spring from the ventricle. That most to the left is the **pulmonary artery**; that apparently next is the **left aorta**; the remaining one is the **right aorta**, which latter is, however, covered close to the heart by the **innominate trunk**, which it gives off. On tracing the aortæ carefully to the ventricle it will be found that the attachment of the left aorta is rather more to the right than that of its fellow, the two crossing close to the heart.

95. The **innominate artery** divides almost immediately into the *right* and *left subclavian*, and *right* and *left carotid arteries*. Dissect these out.

THE ARTERIAL SYSTEM. 43

96. The **subclavians** lie, at first, ventral to the carotids, and, diverging, embrace in the angle between them the reddish **thyroid body**; on being traced, each will be found to give off:

   a. A minute thyroid artery, distributed to the organ of that name.

   b. The *ascending cervical*, which is distributed on the ventral surface of the neck and throat.

   c. Farther on, the subclavian passing forward and outward gives off a short trunk, which soon divides into branches distributed to muscles about the shoulder and pectoral arch.

   d. Then turning dorsally, the subclavian gives off the *superior cervical*, which runs to the side of the neck near its base, and there breaks up.

   e. Next, a branch, which runs to the side of the vertebral column, and there passes in between the first costal plate and the first rib.

   f. Then it gives off the *axillary artery*, which, crossing dorsal to the large cords of the brachial plexus, turns round the *subscapularis muscle*, and becomes the *brachial artery*, which is distributed to the forelimb.

   g. The remainder of the subclavian continues as the *internal mammary artery*, which runs outward to the edge of the carapace, and runs back along it, in the hollow of the marginal plate, to anastomose with the epigastric, which passes forward from the posterior portion of the dorsal aorta (110, *c*).

97. The **carotid arteries**, first dorsal and then external to the subclavians, cross dorsal to the latter and reach

their inner sides ; they then run along the neck to the head. About half way up the neck they come into relation with the *sympathetic* and *pneumogastric nerves*, and the *internal jugular vein*. Trace each carotid, taking great care of the nerves near its upper end, until it enters the skull by a foramen (26, *e*), just in front of the auditory capsule.

98. As each carotid is dissected, the **thymus gland** will be found—a loosely aggregated, yellowish mass, near the base of the neck on each side, and internal to the artery.

99. It will be best to trace back the **pneumogastric nerves** now, as their terminal branches are apt to be injured in dissecting the aortic arches. Just behind the hyoidean apparatus, the carotid artery has the vagus on its outer, and the cervical sympathetic on its inner side. For the present follow the nerves from the skull, beginning about an inch and a half behind the cranium. A short way from the head, the sympathetic comes into close relation with the vagus, so that care is needed to separate them ; finally, about opposite the fifth cervical vertebra, the nerves separate, the sympathetic passing out toward the axilla.

100. Follow the left pneumogastric till it crosses the left aorta ; at this point it gives off the *recurrent laryngeal nerve*, which, turning round the aorta, runs up the neck alongside the gullet. After giving off this branch, the vagus stem breaks up into a leash of branches, some of which run mesially toward the heart, while others may readily be traced along the bronchus to the lung, and to the ventral side of the stomach near the cardia.

THE ARTERIAL SYSTEM. 45

101. The right pneumogastric has a similar distribution; its terminal gastric branches are very long, and cross behind the heart to the dorsal side of the stomach.

102. Divide the fibrous bands which pass between the right lung and the right lobe of the liver, taking great care not to prick the lung. Next push the right liver-lobe over to the left side, and carefully separate the lung from the muscles lying on the vertebral column. Then trace **the right aorta**; it will be seen to arch over the pulmonary vessels and the right bronchus, and to receive a *communicating branch* from the left aorta, near the dorsal side of the visceral cavity and opposite the fourth dorsal vertebra. From the point of union, the *dorsal aorta* will be seen passing backward.

103. Carefully divide, on the left side, the connective tissue uniting the lung to the cardiac end of the stomach, and the lung to the muscles on the dorsal vertebræ, and trace **the left aorta**. Passing round the left bronchus and pulmonary vessels to the dorsal side, it ends by dividing into three main trunks, which are to be traced successively by spreading out the mesentery and pushing aside the viscera.

104. The external division of the left aorta gives off:
 a. The *gastric artery*, which runs a short way on the dorsal aspect of the stomach and then bifurcates to form the *right gastro-epiploic artery*, which runs along the great curvature of the stomach; and the *coronary artery of the stomach*, which runs along the small curvature of that organ.

b. The remainder of this division runs on as the *communicating artery* to join the right aorta.

105. The middle division is the *superior mesenteric;* it breaks up in the mesentery (87) to supply most of the gut, and also gives a branch to the spleen.

106. The third division (*the gastro-duodenal artery*) divides into two main stems, the *superior* and *inferior pancreatico-duodenal arteries*, the former running to the left, and the latter to the right end of the pancreas. Each then courses between duodenum and pancreas to anastomose with the other, giving off branches to the gut and gland on the way.

The *superior pancreatico-duodenal* gives off a branch, the *left gastro-epiploic artery*, which runs along the great curvature of the stomach, and inosculates with the right gastro-epiploic.

107. Now divide the intestine close to the stomach, and again in its rectal portion about two inches in front of the pelvis. Remove the portion of the gut between the two incisions by cutting through the mesentery close to it; while so doing note the large fold, with its concavity to the left, made by the *colon* as it proceeds from the junction with the small intestine, and before turning directly back to the cloaca as the rectum.

108. Spread out the removed portion of the gut; note that the **small intestine** is from five to six times longer than the carapace, and the **large** about one-fourth the length of the small. Then open the small intestine along its whole length, and note its thick muscular coat

THE ARTERIAL SYSTEM. 47

and the longitudinal folding of its mucous membrane. Pass bristles in from the pancreatic and cystic ducts, and observe that each passes obliquely through the intestinal wall.

109. Open the cœcum and colon along their free borders; note the absence of an ileo-colic valve, and the muscular and mucous coats of the tube.

110. The **dorsal aorta** may now be followed back from its point of formation (102). Passing between the kidneys, it ends behind them by dividing into the *right* and the *left common iliacs*. In dissecting its branches, which is a matter of some difficulty, disturb the renal and reproductive organs as little as possible. With care the following branches will be found:

   a. Several small twigs to the kidneys and genital glands.
   b. A pair of branches which run dorsally, and enter the canal on each side of the vertebral column, which is arched over by the necks of the ribs.
   c. A branch (*epigastric artery*) on each side, which runs out dorsal to the kidney; and, reaching the edge of the carapace, runs forward to anastomose with the internal mammary (96, *g*).
   d. A branch on each side, distributed to the pelvic muscles.

111. Each of the **common iliac arteries**, the terminal branches of the dorsal aorta, soon divides into an *internal* and an *external iliac*.

112. The left **internal iliac** almost immediately gives off a large branch (*inferior mesenteric*) to the colon, and the right a much smaller twig to the rectum; the trunks then proceed to the pelvis and are distributed to the penis or clitoris.

113. Each **external iliac** passes outward and backward, and divides into—
    a. A branch distributed to the muscles of the pelvis and the inner aspect of the thigh.
    b. The *sciatic artery*, which supplies most of the lower limb. It takes a more dorsal direction than *a*, and approaches the carapace in the region of the sacral vertebræ, there coming into relation with the *sciatic nerve*. Giving off in this region a small branch, the artery turns outward with the nerve, and the two (passing on the mesial side of the ilium) reach the back of the leg, where they may be found and traced in their further distribution by dividing the skin and subcutaneous tissue along a line drawn from the inner side of the dorsal end of the ilium to the outer side of the knee-joint.

114. Clean and trace **the pulmonary artery**; immediately after springing from the ventricle it divides into a right and left branch. The right passes dorsally to all the other trunks proceeding from the ventricle on the right side of the heart, and is found lying posterior to, and parallel with, the right aorta for some distance; it then crosses ventrally to this vessel and enters the lung. The left pulmonary artery runs forward and outward to its lung.

115. Trace **the pulmonary veins**, one from each lung,

near which organ the vessels lie alongside the bronchus and pulmonary artery; these veins open into the left auricle.

116. Note the small **coronary arteries** and **veins** lying on the surface of the ventricle. The arteries spring from the innominate trunk, and the veins enter the venous sinus.

117. Trace out the system of the two **superior cavæ**; as a preparatory step, remove the skin from the dorsal side of the neck and inflate the vessels from some one of their smaller branches. This inflation will need to be repeated from time to time as the dissection proceeds, as many of the branches when empty are very hard to follow.

118. The superior cava, shading insensibly into the venous sinus at one end, is composed at its other, a short way from the heart, by the junction of three considerable trunks. Dissect them out on the left side.
    a. The inner division (**internal jugular**) runs up the side of the neck, where it is found accompanying the carotid artery and vagus nerve.
    b. The middle and largest division (**subclavian**) may be traced to the forelimb. Near the base of the neck it receives a cervical branch (*external jugular*) which is to be followed up the neck to the neighborhood of the occiput, where it is formed by the union of a branch (*facial vein*) from the mandibular region with a cross branch, which may be followed across the middle line by dividing some muscles pass-

ing from the occiput to the cervical vertebræ, and beneath which it passes to join its fellow. Into this cross branch opens, on each side, a vein from the exterior of the cranium, and also one from inside the skull cavity.

c. Beyond the point where the subclavian vein receives the external jugular, it gets a branch from the regions near the base of the neck on the dorsal aspect; it then continues to the forelimb.

119. The outer division of the superior cava crosses over the armpit, receiving some branches in that region, and then runs, as the **internal mammary**, along the margin of the carapace with the artery of that name (96, g).

120. Before proceeding to study the inferior vena cava, it will be necessary to dissect **the Genito-urinary Apparatus.** Endeavor to get your next specimen (193) of the other sex,* and on it complete the examination of the reproductive organs.

121. If your present specimen be a female, proceed next to section 138; if it be a male, examine the **Male Reproductive Organs;** for this purpose divide the pelvic bones along the median ventral line and push the halves apart; then divide the muscles beneath.

---

* It is not easy to tell the sex of a Slider from external examination; dealers profess to tell by a supposed greater convexity of the posterior part of the carapace in the female. But we have not found this reliable. When a number of students are working together in a laboratory, they will, however, be able to arrange it between them so that all shall get a specimen of each sex.

MALE REPRODUCTIVE ORGANS.

Clean carefully the ventral wall of the thin **cloaca**, back to its external aperture ; meeting on it in the median line will be found a pair of thin, rounded muscles, *retractors of the penis*, which pass at first backward, and then, turning around another muscle, forward and dorsally to their origin. Through the ventral wall of the cloaca the large, dark-colored *penis* will be seen.

122. Pass a pair of forceps through the opening of the extremely distensible cloaca, and dilate the latter by spreading the prongs of the forceps apart. Note the **penis** attached for some way along the ventral side of the cloaca, terminating posteriorly in a free enlargement (*glans*), and having a deep groove along its dorsal surface.

123. Slit the cloaca open along one side, taking care not to injure the penis, and examine the latter ; it is composed of two elongated *corpora cavernosa* placed side by side, and in contact in the median line. Each corpus cavernosum, narrowing as it runs forward, ends anteriorly in a highly vascular dilatation or *bulb*. This is best seen through the ventral side of the cloaca.

124. Make a small hole in a vein which leaves the bulb upon its ventral and outer side ; insert the end of a blow-pipe and inflate the penis ; note its extreme distensibility, and how it becomes erect when distended. The corpora cavernosa at the same time swell up and press against each other, converting the groove on the dorsum of the penis into a canal. Observe, also, the great distention of the glans, and the elevated lips bounding the end of the urethral groove on it ; also, the curvature of the erected organ (owing to the greater elasticity of the tissues upon its dorsal side), which causes it to bend forward and ventrally.

125. Raise the **urinary bladder**, and observe that it ends in a narrow *neck*, which is attached to the front end of the cloaca; make a hole in it and pass a probe from its cavity into the cloaca; pass another probe into the cloaca from the cut end of the rectum (107).

126. Bend the penis forward and fix it; wash the mucus away from inside the cloaca, and find where the probes have entered the latter. That from the bladder appears through an opening at the end of the urethral groove; the rectal opening is dorsal to this.

127. Dorsal and a little to one side of the bladder, opening on each side, are the large apertures of the **accessory bladders.**

128. Replace the penis, turn the bladder back, and fix it in position. Wash out the abdominal cavity with a stream of water, and, filling the carapace with clean water, examine the reproductive glands.

129. **The testes,** a pair of rounded yellow bodies, are placed a short way in front of the pelvis. Each has on it a darkly pigmented, convoluted mass, the *epididymis*, which extends back beyond it.

130. Turning the testes and epididymis toward the middle line, seek a long tube thrown into curves and ending anteriorly in a small dilatation placed dorsal to

MALE REPRODUCTIVE ORGANS.

the posterior end of the lung; posteriorly this tube narrows greatly, but it may be traced along the dorsal side of the epididymis to near the cloaca, where it ends blindly. It appears to be the unaborted *Müllerian duct* of the embryo, and therefore homologous with the oviduct of the female.

131. In the cloaca, a short way external and posterior to the proximal end of the urethral groove, seek the small slit-like aperture of each *vas deferens*. On introducing a probe from the cloaca, the duct may be traced for a considerable distance along the ventral border of the testis.

132. Raise the testicle so as to stretch the band of peritoneum between it and the epididymis. In this band will be seen a number of small *vasa efferentia* passing from the testis to the epididymis. Then dissect away the peritoneum covering the epididymis, and observe that the latter organ is made of a much-coiled tube, with its folds bound together by pigmented connective tissue.

133. Turn the left testis and epididymis to the middle line, and push the posterior end of the lung out of the way. The outer half and the anterior end of the ventral surface of the red kidney will then come into view; and, running along it from before back, close to the point where the connective tissue binding it to the epididymis is attached, a large vein will be seen. Carefully trace the branches which enter this; they are:

    a. From behind, a trunk from the penis, through which the whole set may be conveniently inflated.

b. Entering through a fissure on the outer edge of the kidney, between two of its lobes, the **reni-portal vein.** Trace this back through the muscles of the groin-hollow to the point up to which it has already been dissected (81).

c. A branch entering through another fissure, placed farther forward on the outer margin of the kidney. This vein comes from the lateral regions of the posterior part of the carapace.

d. Farther forward, and in still another fissure, a vein which proceeds from the passage alongside the vertebral centra, which is arched over by the necks of the ribs.

134. Now, replacing the testis and epididymis, find the **posterior vena cava,** lying on the muscles in front of the vertebral column and beside the dorsal aorta. Follow it forward until it enters the right lobe of the liver, and then, cutting away the hepatic tissue, trace it up to the venous sinus, receiving the right hepatic veins on its way, and also the right anterior abdominal vein.

135. Next trace the vein back, turning aside the reproductive organs where they come in the way and also tearing through the bands of peritoneum slinging them. The vessel will be seen to commence opposite the kidneys by the union of several *renal veins,* with branches from the testes and epididymes.

136. Cut away the left testis and epididymis, and clean the ventral surface of the **kidney** of the same side. The gland will be seen to be lobulated, and to present externally the appearance of being made up of a greatly

FEMALE REPRODUCTIVE ORGANS. 55

convoluted tube. From its ventral surface several ducts pass out, and unite to form the *ureter.* Trace this duct to the cloaca, and find its aperture in the latter. It lies at the proximal end of the urethral groove, outside, but close to the aperture of the bladder.

137. Note the bright yellow **adrenal body** lying along the mesial border of the kidney.

138. If your specimen be a male, now pass on to section 143. If it be a female, proceed to study **the Reproductive Organs of the Female.** Remove the ventral side of the pelvis, as directed in section 121. Note the slender homologues of the retractors of the penis, and also the **clitoris,** much resembling a miniature penis. Then open the cloaca on one side (123) and study the clitoris more clearly; observe the *urethral groove* between its two small *corpora cavernosa,* and the relatively large *glans clitoridis* in which it ends.

139. Examine the urinary and accessory bladders (125, 126, 127), which resemble those of the male.

140. Examine the **ovaries:** granular-looking orangered masses, with ova of varying maturity and very different sizes in them.

141. Find the **oviduct.** It is a slightly convoluted tube lying in a broad fold of peritoneum, and opening by a wide mouth into the peritoneal cavity dorsal to the posterior end of the lung. Follow it to the cloaca, where its walls are thicker than elsewhere. Then, on the interior of the cloaca seek two small papillæ, one on either side of the bladder aperture; on these the oviducts open; pass a probe through one. If the oviduct contains no ova it is often hard to find; in that case inflate it from its cloacal termination.

142. Next make the dissections described in sections 133–137 inclusive; reading, where necessary, "ovary," for "testis and epididymis."

143. Remove **the heart** very carefully, dividing the vessels at some distance from it where practicable. Place it on wax in a shallow dish containing dilute alcohol; fix it (by pins passed through the vessels) in various positions as may be most convenient, and carefully make out the points at which, as viewed from the exterior, various vessels enter and leave it (94).

144. Next cut away cautiously the ventral side of the ventricle, and note the incomplete septum attached to this wall and partially cutting off a small right portion (*cavum pulmonale*) from a larger left division of the ventricular cavity. Projecting into the left division is the posterior free end of the auricular septum, which thus incompletely divides the left portion of the ventricle into a right half (*cavum venosum*) which gets blood from the right auricle, and a left (*cavum arteriosum*) which receives blood from the left auricle.

145. Open the pulmonary artery and pass a bristle, guarded at the end with a small knob of sealing-wax, along it to the heart. Though the furthest *left* of the vessels springing from the ventricle when the heart is viewed from outside, the pulmonary artery will be

THE RESPIRATORY ORGANS. 57

found to open into the furthest right subdivision of the ventricle—the *cavum pulmonale*. Pass bristles similarly along the aortæ; both will be found to open into the median division of the ventricle—the *cavum venosum;* the opening of the left aorta being to the right of that of the right aorta. Slit open the arteries, and observe the valves at their mouths.

146. Observe the **Trachea;** note the way it is curved when the neck is retracted, and its division on the right side of the retracted neck into the *right* and *left bronchus*, each of which enters its own lung near the anterior end, alongside the pulmonary artery and vein.

147. Remove the hyoidean apparatus, with the tongue, larynx, and a bit of the trachea attached. To do this, divide the muscles, etc., between the rami of the lower jaw and the anterior horns of the hyoid, so as to open the buccal cavity from its floor; then seize the tongue with forceps and draw it out; note the *stylohyoid ligament* passing from the anterior horn of the hyoid to the base of the skull. Divide it on each side; cut through the trachea about an inch behind the larynx, also any other connecting bands, and remove the mass. Note the eminence at the back of the tongue (4, *c*), and the slit (*aditus laryngis*) in it which leads into the larynx. Separate the firm edges of the slit, and examine the interior of the *larynx*, wiping out mucus, etc., with a little bit of sponge held by a small pair of forceps.

148. Put the removed parts away to macerate in water for a few days; then separate and examine **the**

Larynx and Hyoidean Apparatus; the description of the latter has already been given (57).

- a. The framework of the larynx is composed of three cartilages: a posterior large, circular piece (*crico-thyroid*), and a pair of smaller (*arytenoid*) articulated with its anterior margin. The crico-thyroid cartilage fits closely into the concavity at the back of the body of the hyoid, and is thin and membranous in the middle of its convex anterior surface. The arytenoids bound the sides of the aperture (*aditus laryngis*) which lies at the base of the tongue (4, *c*).

149. Inflate the lungs fully from the cut end of the windpipe, and then ligature the latter to prevent the escape of the air. Note:
- a. The position of **the Lungs,** on either side of the vertebral column, in the hollow of the carapace.
- b. The layer of *pleuro-peritoneum*, which covers them and bends them down to the bones pretty closely in front, but leaves their posterior ends free.
- c. Through the serous membrane, a thin muscular stratum (**diaphragm**) may be seen, forming a sheet all over the lung, but thicker in front than behind and dorsally than ventrally.
- d. Pushing one lung aside, divide the diaphragm close to the vertebral column, and dissect out the organ. As the lung is removed, its enveloping muscle will be seen to arise from the necks of some of the ribs, from the vertebral

THE LUNGS. 59

centra between them, and from a small neighboring area of the costal plates. From these attachments the fibres spread out on all sides to invest the lung.

150. Cut through the bronchus, and entirely remove the lung. Inflate it from the bronchus, and keep it distended by tying the latter; then hang it up for a day or two to dry. When dry divide it into a dorsal and ventral half, and observe its internal structure. Note:

 a. That the bronchus is continued by a large tube running to the posterior end of the lung, but having no cartilages in its walls, except for a very short distance near its commencement.

 b. That from this main *bronchial tube*, others diverge at right angles to its general direction, and end in incompletely separated *air-cells*, which are much smaller in the anterior than in the posterior portion of the lung.

151. Trace the Œsophagus up to the mouth, and, passing a probe down from the latter, note the great distensibility of the upper part of the gullet. Then remove stomach and œsophagus, and, opening them, note the folded mucous membrane lining them; the *rugæ* being most conspicuous in the stomach.

152. Remove carefully the remaining lung, the liver, and the genito-urinary organs, taking care not to injure the dorsal spinal nerves, which are seen running out across the interior of the carapace; then clean the great **Retractor Muscles of the Head and Neck,** of which there are four on each side. Posteriorly they arise without distinct tendons, from the bodies of the dorsal vertebræ from the fourth to the eighth, the one which extends farthest forward having its origin farthest back. This, the longest of the muscles, ends in front by a tendon which is inserted on the posterior part of the skull; the next is inserted on the ventral surface of the anterior outer angle of the centrum of the fifth cervical vertebra; the remaining muscles are similarly fixed to the sixth and seventh cervical vertebræ respectively.

The **Protractors of the Head and Neck,** which run from the carapace and pectoral arch to the neck, may now also be noted.

153. Cut away the retractor muscles; also the ventral three-fourths of the ilium on the side where the **sciatic nerve** has already been seen (113, *b*), and trace the nerve down the limb, and also up to its formation by the union of four trunks (**sciatic plexus**) which appear through the inter-vertebral foramen between the last dorsal and first sacral vertebræ and the three succeeding foramina. The second of these trunks is the largest; the fourth the smallest. This plexus also communicates with the lumbar plexus (154) by a small branch, and gives some small twigs to the pelvis.

154. The nerve-trunks which appear through the two foramina on each side which are next in front of those through which the roots of the sciatic plexus pass out, unite to form the **lumbar plexus,** offshoots of which may be traced to the inner aspect of the thigh.

155. Clear away the muscles, etc., on the anterior

THE SPINAL NERVES.

part of the tail, and note the small **caudal spinal nerves** as they appear through their foramina of exit.

156. Running out across the carapace, note the main branches of the **dorsal spinal nerves**, from the second (which appears between the second and third dorsal vertebræ) to the seventh, inclusive. Each consists of two main divisions which lie side by side.

157. The first dorsal nerve gives off, close to the vertebral column, a branch which enters the brachial plexus (159), and then runs out dorsal to the shoulder-girdle. Parallel to the communicating branch to the brachial plexus, and close to it, is a branch of the main sympathetic chain, which may be noted as a guide in the subsequent dissection of the sympathetic system.

158. There are nine **cervical spinal nerves**, which, except the first, may be found on clearing away the muscles on the sides of the neck. The first (*sub-occipital nerve*) is very small and has no posterior root. To find it flex the head on the spine, open the membrane between occiput and atlas, and with bone forceps cut away the dorsal half of the latter bone. On then gently raising the exposed part of the spinal cord and pushing it to one side, the first cervical nerve will be seen running out between the occipital bone and the atlas. Do not confound it with the spinal accessory (180, *h*), or hypoglossal (180, *i*), which can also be seen; the former runs on into the cranial cavity; the latter passes out through a foramen in the exoccipital bone.

159. The seventh, eighth, ninth, and tenth cervical spinal nerves form the **brachial plexus**, which also gets a branch (157) from the first dorsal nerve. Trace the trunks springing from the plexus to their distribution in the fore limbs.

160. Clear away the muscles still attached to the spinal column in the carapace region and remove what remains of the pelvis, taking care to leave the sacral ribs. Note that the sacrum has a considerable amount of mobility from side to side, the enlarged outer end of the first sacral rib gliding over a smooth elevation on the carapace (15). Such movements, when the parts are in their natural connections, would carry the pelvis with them, so that the pelvic arch is less rigidly fixed than in most vertebrates. Next, remove the pectoral girdle, noting the stout ligament which fixes the dorsal end of each scapula to the carapace.

161. Cut away the centra of all the vertebræ sufficiently to open the neural canal and expose the ventral surface of the *dura mater*. Then slit open this membrane along the middle line, and reflect it to each side, so as to expose the spinal cord; take great care not to injure the roots of the spinal nerves. Note the considerable space between the cord and the dura mater.

162. The **spinal cord**, in its course, will be seen to present two considerable *enlargements*. The *anterior*, or *cervical*, extends from the front of the sixth cervical to the back of the first dorsal vertebra, and gives origin to the roots which enter the brachial plexus. The *posterior*, or *lumbar enlargement*, extends from the ninth dorsal to the first caudal vertebra inclusive, and gives origin to the trunks of the sciatic plexus.

THE SPINAL CORD. 63

The cord is comparatively very slender in the region between the two enlargements, and, in the tail region, gradually tapers off; but it extends to the end of the vertebral column, there being no *cauda equina;* along its ventral aspect the cord presents a well-marked median *anterior fissure.*

163. Carefully raise a bit of the cord in the region of the lumbar enlargement; note that the spinal trunks are formed by the union, outside the neural canal, of a *ventral (anterior)* and *dorsal (posterior)* root, and that each root in turn arises from the cord by a number of small separate bundles. To see the roots well a handlens will be found of use.

164. Examine the roots in the middle dorsal region, and in the neck. In the former they are very small.

165. Carefully trace a pair of roots in the middle dorsal region. They will be found to reach a comparatively large *spinal ganglion* which lies, imbedded in pigmented connective tissue, in one of the hollows found along the vertebral column, between the necks of the ribs. From the outer side of the ganglion proceed the two trunks already referred to (156); they are considerably larger than the roots entering the ganglion.

Next trace out some roots in the anterior cervical region and the cervical and lumbar enlargements, and note the ganglia attached to them.

166. Cutting away the bone more freely, remove parts of the spinal cord, and make sections of it at different levels. Note the well-marked *anterior fissure;*

the conspicuous *gray matter* in the centre in relatively small amount, but with well-marked *cornua* and a distinct *gray commissure*.

On the dorsal aspect of the cord, note the shallow groove indicating the *posterior fissure*, and the furrow on each side of it along which the posterior roots are attached.

167. Now proceed to the **Dissection of the Brain and Cranial Nerves** and the **Sense Organs** connected with them. With a jeweller's saw cut through the soft parts and cranium in a plane parallel to the median antero-posterior vertical plane of the head, and passing in front through the outer margin of the right nostril; carry your incision from above down through the base of the skull, but leave the mandible undivided. The skull will thus be divided into a larger left and a smaller right segment, the former containing the brain, which is exposed on its right side by the section.

163. On the cut surface of the left portion, beginning in front, note:

    a. The *nasal chamber*, lined by brown mucous membrane, which is folded near the opening of the anterior nostril.

    b. Separated from the nasal cavity by a thin plate of bone is a small part of the median portion of the right orbito-temporal fossa, with a bit of the *Harderian gland* (which has been divided by the saw) lying in it anteriorly.

    c. Above and internal to this will be found part of the *superior oblique muscle*.

    d. Close to the membranous upper and posterior

THE BRAIN. 65

boundary of the part of the fossa now under examination, will be found two nerves: the *orbito-nasal* or *ophthalmic*, a branch of the fifth cranial, and the *patheticus* (fourth cranial); the former is the larger.

e. Gently removing the bit of Harderian gland, observe the membranous *inter-orbital septum*, and trace the orbito-nasal nerve forward along it until it enters the nasal chamber. The origin of the superior oblique muscle is above this point of exit, and the fourth nerve may be traced into the muscle near to its origin.

169. Lying above the inter-orbital septum is a prolongation of the cranial cavity, covered in by dura mater, and roofed over by bone. On cutting away the bone above, and the dura-mater on the side, the whole of the right side of the brain will be exposed if the original cut were made in the proper plane, passing behind through the right margin of the foramen magnum. If the cut has taken a too external direction behind, carefully break the bone away from time to time as may be necessary to observe the parts described below.

170. On the side of **the brain**, note:
 a. The right *olfactory lobe*, reaching in front to the posterior boundary of the nostril and marked off behind by a shallow furrow from—
 b. The right *cerebral hemisphere*, which is separated behind by a deeper furrow from—
 c. The *mid-brain* or *mesencephalon*, behind which comes—
 d. The *cerebellum*.

- *e.* Ventral to the cerebellum, and reaching anteriorly to the mid-brain, is the *medulla oblongata* or *myeloncephalon*, which is seen to be continuous posteriorly with the spinal cord or *myelon*.
- *f.* The medulla and the ventral parts of the mid-brain are white; the other portions of the brain now exposed are gray.

171. Carefully raise the cerebral hemisphere so as to expose beneath its posterior portion the *thalamencephalon*, and under this note the large, white *optic tract* running forward from the mid-brain to join its fellow in the *optic chiasma;* from the latter, trace the right *optic nerve* forward to the point where it has been divided. It will be seen to turn out into the orbital fossa, just behind the inter-orbital septum; and, as soon as it enters the fossa, to cross out beneath the *ophthalmic nerve*, which runs between it and the *superior rectus muscle* near the origin of the latter.

172. About the point where cerebellum, medulla oblongata, and mid-brain meet, the posterior end of the small *fourth nerve* will be seen; trace it forward. Next, gently raise the mid-brain from the base of the skull, and note the *third cranial nerve* (*motor oculi*) running from it to the orbit, which it enters somewhat posterior and external to the optic trunk.

173. On the small right hand portion of the skull (167) will be seen:
- *a.* A much larger portion of the nasal chamber than that which was found on the left section.

b. Behind this the main portion of the right orbital fossa, containing the eyeball and most of its muscles, and the outer part of the optic nerve attached to the globe. Entering the orbit behind and running forward near its roof will be seen a slender branch of the third nerve.

174. Behind the orbit is a thin plate of the parietal bone (45), separating the cranial cavity from the temporal fossa. On its inner side will be found the sixth nerve, running forward to the orbit, through which cavity it may be traced to the external rectus muscle. On the inner surface of the bone is also a *fossa*, or depression, opposite the back part of the cerebral hemisphere and the front of the mid-brain, and placed between the parietal and proötic bones. Pull away the dura mater which covers this pit, and the large *Gasserian ganglion* will be exposed, with the sensory root of the fifth cranial nerve entering it. Proceeding forward from the ganglion, will be found the stump of the orbitonasal nerve, the peripheral part of which has already been seen (168, *d*).

175. Posterior to the fifth nerve, the *seventh* of the right side will be seen, entering a foramen in the cranial wall; and, close behind it, the smaller *eighth* (*auditory*) nerve, entering the auditory capsule. Above its point of entry the inner wall of the labyrinth is cartilaginous.

176. Still farther back is the foramen of exit for the tenth and eleventh nerves (*pneumogastric* and *spinal accessory*).

177. On the right portion of the divided head, **the Ear** is to be dissected either now or after completing the dissections described in paragraphs 179-192, inclusive. For this purpose—

    a. Remove the skin over the tympanum (3, *f*); note the white, opaque, fibrous ring surrounding the aperture, and the circular cartilaginous plate which nearly fills it up, and is united to its margin by a delicate membrane.

    b. Cut away the uniting membrane, and turn the cartilage (which is the outer expansion of the columella, 26, *d*) to one side; the rod-like portion of the columella will be seen running in from it. Cut off the outer end of the columella, and examine again the outer division of the tympanum (27, *a*) now lined by mucous membrane; the columella traverses it and disappears on its inner side through a narrow passage. Beneath and behind the columella is a slit bounded by a pair of white prominent lips; this is the tympanic opening of the *Eustachian tube;* pass a guarded bristle along it to its oral end, which will be found to open on the side of the buccal cavity, just behind the prominence caused by the powerful muscles which run from the temporal fossa to the mandible.

    c. Carefully break away the floor of the tympanum, and trace the columella through its fibrous canal to the inner dilatation of the tympanum (26, *d*), where it ends in a circular cartilaginous plate, smaller than the outer one, fixed in the foramen ovale.

THE INTERNAL EAR.

178. Break away the base of the skull internal to the tympanum, so as to open the vestibule, which, if the saw was held in the proper plane while dividing the skull, will not have been injured. The vestibule has three semicircular canals opening into it, and in front is a conical prolongation of its cavity, which represents a rudimentary *cochlea*.

   a. The *semicircular canals* are not very easy to trace on a fresh specimen. If they are not recognized, take first a dried macerated skull, and open the vestibule from its base; clear away the membrane lining it, and note the recesses of its wall: an anterior at the base of the cochlea, a superior, and a posterior; into these semicircular canals open. The *anterior* and *posterior vertical canals* open into the upper recess by their non-ampullary ends; a fine bristle pushed forward from this recess passes on through the anterior vertical canal and appears again, through its ampullary enlargement, in the anterior recess. From the posterior recess the *posterior vertical* and *horizontal canals* proceed; the former here has its ampulla, and through it a bristle may be passed to the superior recess. Push another bristle forward from the opening of the horizontal canal; it will reappear in the anterior recess, where the ampulla of this canal opens.

   b. Having now found the position of the openings, trace out the canals on your fresh specimen.

179. Returning to the left side of your section of the

head, break away the bone on the roof of the skull, taking care not to injure the auditory capsule, and note from before back, on the dorsal aspect of the brain :
- *a.* The olfactory lobes.
- *b.* The cerebral hemispheres, or *prosencephalic lobes*, with the *pineal gland* lying in the angle where they diverge behind.
- *c.* The *corpora bigemina*, forming the dorsal part of the mid-brain.
- *d.* The unpaired *cerebellum ;* and, behind it—
- *e.* The *medulla oblongata*, covered by a highly vascular plug of *pia mater*, forming a *choroid plexus*. On removing this, the *fourth cerebral ventricle* will be partly exposed on the dorsal aspect of the medulla. Raise the posterior free border of the cerebellum, and note that the fourth ventricle is continued forward beneath it.

180. Carefully tilt the brain over toward the right, and study the origin of the various **Cranial Nerves** on the left side.
- *a.* The *olfactory* (*first cranial nerve*), proceeding from the end of the olfactory lobes.
- *b.* The *optic* or *second cranial nerve* (171).
- *c.* The *third nerve* (*oculo-motor*) arising from the ventral aspect (*crus cerebri*) of the mid-brain.
- *d.* The *fourth nerve* (*patheticus*) will be first found on the dorsal side of the crus cerebri, close to the cerebellum ; thence it passes forward and downward till it reaches the base of the skull, on which it runs forward parallel and external to the third nerve. The three preceding

THE CRANIAL NERVES.   71

nerves all disappear through the dura mater closing in the skull cavity in front, and separating it from the orbital cavity.

e. Carefully separate the soft parts beneath from the bony roof of the temporal fossa and orbit, and chip away the bone. Note the large *fifth nerve* (*trigeminal*) arising from the anterior part of the side of the medulla, and passing through the dura mater; also the *Gasserian ganglion* (174), which may be exposed by removing the membrane over it.

f. Posterior to the fifth nerve, the *seventh* (*facial*) and *eighth* (*auditory*) rise together from the side of the medulla; they soon diverge and pass out by separate foramina, that of the seventh being anterior to the cartilaginous inner side of the labyrinth.

g. The *glosso-pharyngeal nerve* (*ninth cranial*) arises from the medulla, and leaves the skull by a foramen in the opisthotic bone immediately behind the cartilaginous portion of the inner wall of the labyrinth.

h. Immediately behind the glosso-pharyngeal a number of nerve-roots arise from the side of the medulla; the anterior unite to form the tenth nerve (*pneumogastric*); the posterior (reinforced by filaments from the anterior end of the spinal cord) form the eleventh nerve, or *spinal accessory.* The two leave the skull by a large common foramen, between the exoccipital and opisthotic bones.

i. The *sixth* and *twelfth cranial nerves (abducens and hypoglossal* respectively) have foramina of exit nearer the median line than any of the above, so that the brain must be turned farther over in order to see them. The sixth arises from the ventral aspect of the medulla near its anterior end, and, after a short course, enters a canal in the basisphenoid bone; farther forward it reappears in the cranial cavity beneath the anterior clinoid process and passes on to the region where it has been already described (174). The *twelfth nerve* arises from the ventral aspect of the medulla near its posterior end, and passes out through one of the condyloid foramina.

181. Remove the brain (dividing the nerves close to their origins with sharp scissors) and place it aside in alcohol for future examination; then trace out the branches proceeding from the left Gasserian ganglion. On cleaning away the dura mater, which covers the inner surface of the descending plate of the parietal bone and of the alisphenoid, the *ophthalmic branch* of the fifth nerve will be found. Trace it forward to the orbit; it enters this a little external and posterior to the optic nerve, which will be recognized passing in beneath the superior rectus muscle. In the orbit the ophthalmic crosses under the muscle and appears again on its inner side, and here has the fourth nerve lying close along it.

182. Follow the fourth nerve forward till it enters the *superior oblique muscle*, and then trace out the origin and insertion of the latter; it arises from the frontal bone immediately in front of the membranous inter-orbital septum, and is inserted upon the upper suface of the eyeball.

183. Divide and reflect the superior rectus, looking for the small branch of the third nerve which enters it, and then trace the ophthalmic nerve through the orbit, pushing aside other parts which come in the way, but not dividing them.

> *a.* Soon after it enters the orbit, the ophthalmic trunk divides into a larger and smaller branch. The former crosses above the optic nerve and beneath the superior oblique muscle ; in front of the orbit it passes over the *Harderian gland*, and, on dividing and removing the superior oblique muscle, may be traced forward between the gland and the inner boundary of the orbit till it pierces the latter to enter the nasal chamber.
>
> *b.* The smaller or ventral division of the ophthalmic trunk, turns mesially below the optic nerve, and there enters the *ophthalmic ganglion*. The ganglion is also joined by a considerable branch from the *third nerve*, which lies here to the inner side of the ophthalmic branch. From the ganglion a number of *ciliary nerves* run forward to the eyeball. They may be traced by raising and turning forward the optic trunk.

184. As the ophthalmic nerve enters the eyeball it lies on the *external rectus muscle*, which arises from the hollow on the inner surface of the base of the skull under the anterior clinoid processes, and passes forward to be inserted on the outer side of the eyeball. Carefully separate the outer surface of the posterior portion of this muscle from the side of the skull against which it lies, and note the sixth nerve entering it. Trace the nerve back till it disappears in the bone (180, *i*).

185. Clean the *internal rectus muscle* which passes from the posterior part of the inter-orbital septum to the inner side of the eyeball. On its course it crosses over the Harderian gland. Detach the muscle from the eyeball, carefully turn it back, and note the branch of the third nerve which enters its under surface.

186. Note the *lachrymal gland* on the outer side of the eyeball, and then, separating it and the Harderian gland from the globe, expose the *pyramidalis muscle*. Its fibres arise from the inner side of the eyeball, arch over the optic nerve, and then turn down and around the latter and the external rectus to be inserted into the lower eyelid and the nictitating membrane.

187. Remove the lachrymal gland; raise the eyeball and note the *inferior oblique muscle*, placed transversely near the front of the orbit; it arises from the lower part of the inner wall of the orbit, and is inserted upon the outer side of the eyeball. As the eyeball is being raised note more closely the nictitating lid, and the fibrous prolongation from its lower margin to the pyramidalis muscle, which will now be better seen than previously.

188. Turn the eyeball back (dividing any fibrous bands that interfere) so as to expose its under surface, with the insertion of the *inferior rectus muscle* upon it.

THE CRANIAL NERVES. 75

Then trace the muscle back below the optic nerve to its origin from the base of the skull beneath the anterior clinoid processes.

189. Pushing the eyeball aside, trace forward below it the branch of the third nerve which enters the inferior oblique muscle near its origin. Then remove the eyeball and its muscles entirely.

190. Seek in the loose tissue covering the floor of the orbit for the **superior maxillary nerve**, a branch of the fifth cranial. Having found it, cut away the bony arch covering in the temporal fossa, and trace the nerve back through the mass of muscles occupying that region of the skull, until it reaches its foramen of exit (27, *b*); then, carefully cutting the bone away as required, trace the trunk to the Gasserian ganglion. Immediately after passing out of the skull, the nerve gives off a branch which enters the muscles lying around it. It then runs forward to the orbit, which it enters at its inner side and close to its floor, and there divides into two branches. The smaller of these passes forward and outward, and disappears through a foramen in the anterior region of the floor of the orbit. The larger branch runs forward in a more internal course and passes through the membranous region (29) of the orbital floor to enter the nasal cavity.

191. The **inferior maxillary nerve** leaves the skull by the same foramen as the superior, and may there be readily found on the ventral side of the latter and traced back to the ganglion. After getting outside the skull it gives off several small branches to the neighboring muscles, and a somewhat larger one, which passes forward and enters the *temporal muscles* near the articulation of the mandible.

    a. The remaining principal portion of the nerve then runs downward and forward to the slit in the upper side of the ramus of the lower jaw (53), and divides into two branches, which enter it.

    b. The larger anterior branch runs forward in the mandible, giving off branches in it, and appears again on its inner side, just above Meckel's cartilage (54, *a*), where that first comes into view on the mesial side of the ramus of the mandible. It is then distributed to parts in the floor of the mouth.

    c. The posterior division of the inferior maxillary nerve runs down nearly vertically through the ramus of the mandible and reappears on its inner side, near the lower border of the bone, through a special foramen (53), and is then distributed to the neighboring soft parts.

192. Trace back the sixth nerve, from the part already seen in the orbit till it enters the basisphenoid, and then follow it, by breaking away the bone with great care. The nerve, while in its bony channel, will be found to receive a branch from the Gasserian ganglion, which passes in through a minute bony canal, and a small ganglion is placed at the point of junction. From this ganglion three branches arise. Trace them out.

193. Taking now a fresh terrapin, chloroform and open it in the manner already described (77).

Extend the neck, divide the skin along the ventral median line, and reflect it. Remove the heart and alimentary canal, except the gullet, but leave the lungs and air-passages and the œsophagus undisturbed; cut away the ventral half of the pectoral girdle, but do not injure the brachial plexus (159). Carefully remove the skin covering the hyoidean apparatus between the rami of the mandible, and treat similarly the superficial muscular layer beneath it.

194. The anterior hyoid cornu will now readily be felt through the muscles which lie on it, running parallel to the mandible on each side. At the dorsal end of this cornu, and near its posterior border, will be found three nerve-trunks, which are now to be carefully dissected to their peripheral distribution.

   a. One of these is the **hypoglossal**. It passes forward and divides into an external branch, distributed to the muscles on the anterior side of the cornu, and a lingual division which may be traced on to the tongue-muscles, giving off a few filaments on its way.

   b. Near the hypoglossal, at the point where it first appears in the above dissection, but lying closer to the anterior hyoid cornu, the **glossopharyngeal nerve** will be found; trace it on to its distribution in the space between the two hyoid cornua; one especially prominent branch runs along the posterior border of the anterior horn.

   c. Still another nerve will be found near the proximal end of the cornu, and close against its posterior border; it is a branch of the

seventh nerve, and turning around the cornu may be traced into a muscle which passes from that to the proximal end of the mandible on the same side.

195. Divide the hyoidean apparatus, the larynx, and the ventral walls of the pharynx and gullet in the median line, and turn the halves outward so as to expose the base of the skull, covered by the mucous membrane of the roof of the mouth and pharynx. Then proceed to trace the nerves described below, taking the greatest care not to injure the rest while following one of them.

196. Find the **pneumogastric** in the neck (99), where it has already been seen with the carotid artery and the cervical sympathetic accompanying it. Trace all three from that point to the base of the skull, where the artery (crossed on its course by several nerves, which must not be cut) passes out of sight, entering the canal which opens just behind the proximal end of the quadrate-bone (26, *e*). Then cut away the artery, disentangling it from the nerves around.

197. Next trace the pneumogastric nerve up to the skull (for its peripheral distribution see section 100). It will be seen to enter a ganglion (**superior cervical ganglion** of the sympathetic) which lies on the base of the skull, immediately posterior to the cartilaginous floor of the tympanum.

198. Next trace back the **hypoglossal** from the point where it has already been seen (194, *a*); it crosses on

THE CRANIAL NERVES. 79

the dorsal side of the pneumogastric, and is attached to it at that point. Beyond this follow it till it enters its foramen in the occipital bone (25, *b*).

199. Follow the **sympathetic trunk** along the upper half of the neck. Near the skull it separates from the pneumogastric, and lies to its outer side; crossing ventrally to the hypoglossal, it divides into an inner and outer branch. The former enters the ganglion (197); the latter proceeds alongside the carotid, and disappears through a foramen in the base of the skull (26, *f*) placed just internal to that through which the carotid disappears.

200. A short way anterior and external to the superior cervical ganglion, the **seventh nerve** will be found passing out from its foramen immediately in front of the opening of the carotid canal; it then runs to the proximal end of the anterior cornu of the hyoid. In its course it presents a small ganglion, from which two branches procced; one is that already seen curling around the hyoid cornu to enter the digastric muscle (194, *c*); the other is distributed along the posterior border of the mandible.

201. Trace the glosso-pharyngeal nerve (194, *b*) toward the skull; it will be found to enter the superior cervical ganglion.

202. Two small branches in this region still remain for description: one runs back from the superior cervical ganglion to the gullet, and is probably to be regarded as an *œsophageal branch* of the pneumogastric;

the other (*descendens noni*) arises from the hypoglossal, close to its foramen of exit, and runs back among the muscles of the neck.

203. Repeat on the other side the dissections given in sections 196-202, inclusive, and then, with a pair of bone forceps break away the base of the skull in the middle line, so as to expose the ventral aspect of the medulla oblongata; with care this can be done without injuring the nerves above described, which are then to be followed into the cranial cavity, bits of bone being broken away as may be necessary, with strong scissors or a stout needle mounted on a handle. The hypoglossal may thus be followed to its superficial origin (180, *i*); the pneumogastric, spinal accessory, and glosso-pharyngeal traced from the superior cervical ganglion to the brain; and the seventh nerve followed through the bone in front of the auditory capsule to the cranial cavity. Then remove the eyeballs, and cut off the head and the upper part of the neck and place aside in strong alcohol to harden the brain.

204. To study the **Anatomy of the Eyeball**, clean away the muscles, etc., from its exterior. Note:
    a. The transparent *cornea* in front, forming a segment of a smaller sphere than the semi-opaque *sclerotic* behind, which has a bluish tint from the choroidal pigment seen through it. In the anterior part of the sclerotic bony plates may be felt imbedded in it, forming a ring just where it meets the cornea. The *conjunctiva* on the front of sclerotic contains much dark pigment.
    b. Cut away the cornea; note the *anterior cham-*

THE EYE.

ber of the eye between it and the front of the lens, and filled with the *aqueous humor*.

– Also the brown and gold *iris*, projecting into the anterior chamber, and overlapping the margins of the lens; it surrounds the circular *pupil*.

c. The *crystalline lens* may now be exposed on its anterior aspect by fastening the eye, with its anterior side upward, by pins to a layer of wax in the bottom of a dish, covering it with water, and then cutting away the iris. Next remove the lens, and note its bi-convex form.

d. After removing the lens, the *vitreous humor* will be seen.

e. Take away carefully the vitreous humor, and observe the *retina*, with its delicate vessels, behind it. Detach the retina from—

f. The pigmented *choroid* outside it.

g. As the retina is raised note the point where the optic nerve is attached to it.

h. Raise the choroid, and the inner surface of the sclerotic coat will come into view behind it.

205. After twenty-four hours, proceed to break away the remainder of the base of the skull, and also the bodies of two or three of the anterior cervical vertebræ; take care of the various nerves attached to the brain and spinal cord. As the bone below the interorbital septum is removed, note above it a cartilaginous rod (*trabecula cranii*, or naso-ethmoidal cartilage) which passes forward, and between the nose-chambers expands into a vertically placed cartilaginous lamella, the representative of the ethmoid. This may not have been seen hitherto, as it is apt to be removed in preparing the skeleton (28). Cut it away, with those orbital muscles which arise from the floor of the brain cavity, and expose the optic nerves.

206. Clear away the dura mater which closely invests the exposed ventral part of the cord and of the mid-brain. Then study the general outline of **the brain** as now seen from its ventral aspect. Note:

- *a.* In front, the olfactory lobes (170).
- *b.* Behind them, in the middle line, the *optic commissure*, from which the *optic nerves* diverge anteriorly and the *optic tracts* behind.
- *c.* Between the optic tracts, the *pituitary body* or *hypophysis cerebri*.
- *d.* Meeting at the base of the olfactory lobes in front, but diverging behind, so as to embrace between them the pituitary body and optic tracts, the large *prosencephalic lobes* (*cerebral hemispheres*).
- *e.* Between these, bearing the *infundibulum*, or stalk of the pituitary body, and bounded by the optic commissure in front, is the ventral aspect of the *thalamencephalon*.
- *f.* On the sides of the thalamencephalon, and passing dorsally beneath the optic tracts, are the *crura cerebri*, forming the ventral portion of the mid-brain.
- *g.* Behind the mid-brain are the *anterior pyramids* of the *medulla oblongata*, with a furrow between them, continuous behind with the anterior fissure of the spinal cord.
- *h.* Dorsal to the medulla, and projecting slightly

THE BRAIN. 83

beyond it on each side, will be seen a portion of *the cerebellum.*

207. Remove the brain entirely from the skull cavity, breaking away the bones as may be required. Note that, on its dorsal aspect, the dura mater adheres closely to it as far back as the prosencephalic lobes extend, and that so far there is no separate periosteum for the inner surfaces of the cranial bones. Farther back, on the contrary, the membrane splits into two layers, one of which adheres to the bones as a thin membrane while the other invests the brain as a separate dura mater, and between the two there is a considerable space.

208. Remove the dura mater, the pineal gland, and the processes of the meninges which dip between the main subdivisions of the brain, and study the parts of the encephalon carefully (170–179). Now that it is removed from the skull and hardened, it can be better examined than before. Note:
  a. The *olfactory lobes (rhinencephalon),* which are separated by a shallow groove from one another and from the prosencephalon.
  b. Beginning on the dorsal side, push apart the prosencephalic lobes where they lie in contact in the middle line, until the optic commissure comes into view; note the absence of any corpus callosum uniting them.
  c. The posterior end of each cerebral hemisphere (*prosencephalic lobe*) is free, and diverges from the median plane so as to pass outside the mesencephalon, to the posterior border of

which it nearly extends; on its dorsal aspect the mesencephalon is not overlapped at all by the prosencephalon.

d. In the angle between the mid- and fore-brains, the small *optic thalami* are seen touching one another in the median line.

e. The *mesencephalic* or *optic lobes* (*corpora bigemina*), much smaller than the prosencephalic, are paired hemispherical eminences, white on the surface, and placed on the dorsal side of the brain.

f. Behind the corpora bigemina, and rather larger than both together, is the *cerebellum*, only attached to the rest of the brain by its anterior margin, and overhanging the fore part of the *medulla oblongata*. There is a slight notch in its posterior border, from which a shallow, wide median furrow proceeds forward some way on its dorsal aspect.

g. Raising the hinder part of the cerebellum, note the medulla oblongata, with the *fourth ventricle* on its dorsal side, tapering off posteriorly into the *calamus scriptorius*. Note the two parallel cords of white nerve-substance lying side by side along the floor of the fourth ventricle; they are continuations of the anterior columns of the spinal cord.

h. On the side of the brain, note the relations of the various parts already described, and trace the white band of the optic tract up to the mesencephalic lobe.

i. Pushing the posterior free end of the prosencephalic lobe out of the way on one side, divide the optic tract, and trace beneath it the *crus cerebri* passing up into the prosencephalic lobe.

209. Carefully push apart those parts of the brain between which there is a median fissure, and then, with a sharp, thin-bladed knife divide the whole organ, with the piece of spinal cord attached to it, in the median dorso-ventral plane. On the inner surface of the sections note:
- *a.* The continuity of the fourth ventricle with the central canal of the spinal cord.
- *b.* The form of the cut edge of the cerebellum, as now exposed, which shows that that organ is a thin arched lamina.
- *c.* The passage (*aqueduct of Sylvius*) from the front end of the fourth ventricle to the third, and the communications of this passage with large cavities in the interior of the optic lobes.
- *d.* The *third ventricle*, bounded on each side posteriorly by the flat inner surface of an optic thalamus. Farther forward will be found on each side an aperture (foramen of Monro) leading into the *lateral* (first and second) *cerebral ventricles*, one beneath each prosencephalic lobe.

210. Make cross sections of one half of the brain at different points to see the relations of the parts and the distribution of white and gray matter.

211. Now dissect the reproductive organs of your present specimen (sections 121–137, or 138–141, according to sex). Do not make the dissections of the kidneys or of the veins running to them, or of the inferior cava, as these parts are alike in both sexes and the sympathetic is apt to be injured in examining them.

212. Dissect out the **Sympathetic Chain** on one side. Find the main cord in the neck, where it separates from the pneumogastric (99). Following it, it will be seen to pass outward and backward, and to have on it a large ganglion (**middle cervical**), from which several branches proceed. One of them is larger than the rest, and forms obviously the backward continuation of the main trunk; it proceeds toward the axilla. Amputate the fore limb at the middle of the humerus, and carefully expose the brachial plexus. The sympathetic cord runs toward this, and, before reaching it, presents a third small ganglion (**inferior cervical**); crossing the plexus it then enters a large fusiform ganglion (**ganglion stellatum**) placed on the ventral side of the most posterior of the main trunks of the brachial plexus.

213. Apparently proceeding from the ganglion is a small nerve, which runs to the ventral aspect of the lung and there ends in the muscular layer covering this organ. The nerve probably gets filaments from the ganglion, but careful dissection will show that it is mainly derived from the trunk of the brachial plexus on which the ganglion lies.

214. From the posterior end of the *ganglion stellatum* two filaments run back, one on each side of the subclavian artery. They enter a ganglion which lies just anterior to the first dorsal nerve, and is connected with it by a communicating branch. From this ganglion the sympathetic cord runs back, crossing the neck of the second rib ; and, in the hollow between this and the third, lies on and adheres to the ganglion of the second dorsal spinal nerve (165).

215. Thence the trunk continues to the neck of the third rib, and on it divides into an external and larger branch which enters the muscular layer of the lung, and a smaller one, which continues the direction of the main trunk toward the next spinal ganglion, to which it is not directly adherent, but just beyond which it receives a large communicating branch from this ganglion.

216. The sympathetic cord, still running posteriorly, divides again into a larger outer division, which runs to the lung musculature, and a smaller inner, which passes by the fourth spinal ganglion and gets a communicating branch from it, a small ganglion being found at point of union. The common cord thus formed gives a slender *outer* branch to the lung, and an *inner*, which runs back and gets a branch from the fifth dorsal spinal ganglion, there being a ganglionic enlargement at the point of junction.

217. From the last ganglion three branches run back. One (outer) passes to the dorsal aspect of the kidney, running in the peritoneum; one (internal) enters the tissues about the dorsal aorta; the median and largest continues the chain back, and passes toward the flat surface of the kidney, which lies against the vertebral column. Before reaching this, it gives off a large outer branch, distributed to the kidney and genital glands, while a slender inner filament runs back on the median surface of the kidney and there gets the communicating branches from the sixth and seventh dorsal spinal ganglia, there being a well-marked sympathetic ganglion, which gives off branches to neighboring parts, at the point of union. From this ganglion the sympathetic chain passes to another which lies on the same surface of the kidney farther back, and receives a long, communicating filament from the eighth dorsal spinal ganglion. From the most posterior ganglion on the kidney the main sympathetic cord runs to another, placed a short way posterior to that organ, and getting a communicating branch from the ninth dorsal spinal ganglion. The sympathetic trunk then runs back, and becomes connected with the first trunk of the sciatic plexus close to its ganglion.

218. In the caudal region a longitudinal nerve-cord unites the spinal nerves, and is, perhaps, a continuation of the sympathetic chain.

END.

INDEX.

ABDOMINAL viscera, 39
Abducens (fourth cranial) nerve, 67, 71, 76
Accessorius (eleventh cranial) nerve, 71, 76
Accessory bladders, 52
Aditus laryngis, 8, 57
Adrenal body, 55
Air-cells of lungs, 59
Alisphenoid bone, 25, 29
Angular bone, 27
Anterior abdominal veins, 36, 40
Anterior clinoid process, 24
Anterior nares, 7
Aorta, dorsal, 45, 47; left, 42, 45, 57; right, 42, 45, 57
Apparatus, hyoidean, 7, 30
Aqueduct of Sylvius, 85
Aqueous humor, 81
Arch, pectoral, 31; pelvic, 33
Arches, neural, of dorsal vertebræ, 14
Artery, ascending cervical, 43; axillary, 43; brachial, 43; common carotid, 43, 44, 78; coronary, of stomach, 45; coronary, of heart, 48; epigastric, 47; external carotid, 19, 20, 43; gastric, 45; gastro-duodenal, 46; gastro-epiploic, 45, 46; genital, 47; iliac, 47; inferior mesenteric, 48; innominate, 42; internal carotid, 20; internal mammary, 43; pancreatico-duodenal, 46; pulmonary, 42, 48, 56; renal, 47; sciatic, 48; subclavian, 42; superior cervical, 43; superior mesenteric, 41, 46; thyroid, 43
Articular bone, 27
Arytenoid cartilage, 58

Astragalus, 34
Atlas vertebra, 17
Attrahens pelvis muscle, 36
Auditory (eighth cranial) nerve, 67
Auditory organ, 68
Auricles of heart, 39, 41
Axis vertebra, 17

BASE of skull, 19
Basi-occipital bone, 24, 28
Basi-sphenoid bone, 24, 28
Bladder, accessory, 52; gall, 39; urinary, 38, 40, 52
Bones of carapace, 11; of fore-limb, 31; of hind-limb, 34; of mandible, 26; of pectoral arch, 31; of pelvic arch, 33; of plastron, 16; of skull, 22
Bony skeleton, 11
Brachial artery, 43; plexus, 61
Brain, 64, 70, 82
Bronchus, 59
Buccal cavity, 8

CÆCUM, 40, 47
Calamus scriptorius, 84
Calcaneum, 34
Canals, semicircular, 69
Carapace, 6, 11
Carotid arteries, 19, 20, 43, 44, 78
Carpalia, 33
Carpus, 32
Cartilage, arytenoid, 58; cricothyroid, 58; Meckel's, 26; naso-ethmoidal, 21
Caudal nerves, 61; vein, 38; vertebræ, 18

INDEX.

Cavum arteriosum, 56; pulmonale, 56; venosum, 56
Centra of dorsal vertebræ, 12
Central plates, 10
Centro-lateral plates, 10
Cerebellum, 65, 70, 84
Cerebral hemispheres, 65, 70, 82, 83
Cerebral ventricles, 70, 84, 85
Cervical enlargement, 62; spinal nerves, 61
Cervical vertebræ, 16
Characters, zoölogical, of Sauropsida, 1
Chelonia, characters of, 4
Choroid coat of eye, 81; plexus, 70
Ciliary nerves, 73
Claws, 9, 10
Clavicle, 16
Clitoris, 55
Cloaca, 10, 51
Cochlea, 69
Colon, 46, 47
Columella auris, 20, 68
Common carotid, 43, 44, 78; iliac, 47
Condyle, occipital, 18
Conjunctiva, 80
Cord, spinal, 62
Cornea, 80
Coronary artery of heart, 49; of stomach, 45; vein, 49
Coronoid bone, 27; process, 26
Corpora bigemina, 70, 84; cavernosa, 51, 55
Costal plates, 12, 15
Cranial nerves, 70
Cricothyroid cartilage, 58
Crus, 9, 34
Crus cerebri, 70, 82
Crystalline lens, 81
Cuneiform bone, 32
Cystic duct, 41

DENTARY bone, 27

Dermic bones of carapace, 11; of plastron, 16
Descendens noni, 79
Diaphragm, 58
Digits of manus, 9, 33; of pes, 9, 34
Dorsal aorta, 45, 48; spinal nerves, 61; vertebræ, 11
Duct, cystic, 41; of Müller, 53; pancreatic, 41

Duodenum, 49
Dura mater, 62, 83

EAR, 68

Eighth cranial (auditory) nerve, 67
Eleventh cranial (spinal accessory) nerve, 67, 71
Epicoracoid, 31
Epidermic exoskeleton, 10
Epididymis, 52
Epigastric artery, 47
Epiotic bone, 23
Epiplastron, 16
Eustachian tube, 8, 68
Exoccipital, 24, 28
Exoskeleton, dermic, 11; epidermic, 10
External carotid, 19; characters, 6; iliac, 48; jugular, 49; rectus, 73
Eyeball, dissection of, 80
Eyelids, 7

FACIAL (seventh cranial) nerve, 67, 71, 77; vein, 49
Female reproductive organs, 55
Femoral vein, 37
Femur, 34
Fibula, 34
Fifth cranial (trigeminal) nerve, 67, 71
First cranial (olfactory) nerve, 70
Fissures of spinal cord, 63
Foramen lacerum posterius, 19, 28; magnum, 18; obturator, 34; of Monro, 85
Fore-limbs, 9, 31
Fossa, glenoid, 31; orbito-temporal, 18, 20
Fourth cranial (pathetic) nerve, 65, 66, 70; ventricle, 70, 84
Frontal bone, 25

GALL bladder, 39

Ganglia, Gasserian, 67, 71; inferior cervical, 86; middle cervical, 86; ophthalmic, 73; spinal, 63; superior cervical, 77; stellate, 86
Gasserian ganglion, 67, 71
Gastric artery, 45

INDEX. 91

Gastro-duodenal artery, 46
Gastro-epiploic artery, 45, 46
Genital arteries, 47; organs, female, 55; male, 50
Genito-urinary organs, 50, 55
Gland, Harderian, 65, 66
Glans penis, 51; clitoridis, 55
Glenoid fossa, 31
Glosso-pharyngeal (ninth cranial) nerve, 71, 77, 79
Gullet, 59

HARDERIAN gland, 66

Heart, 39, 41, 56
Hemisphere, cerebral, 65, 70
Hepatic vein, 42
Hind-limbs, 9, 34
Humerus, 32
Hyoidean apparatus, 7, 30, 58
Hyoplastron, 16
Hypoglossal (twelfth cranial) nerve, 71, 77, 78
Hypophysis cerebri, 82
Hypoplastron, 16

ILIAC bone, 33; arteries, 47

Ilium, 33
Inferior cervical ganglion, 86; maxillary nerve, 75; oblique muscle, 74; pancreatico-duodenal artery, 46; rectus muscle, 74; vena cava, 54
Infundibulum, 82
Innominate artery, 42
Interior of skull, 28
Internal ear, 69; iliac artery, 47; jugular vein, 44, 49; mammary artery, 43; mammary vein, 50; rectus muscle, 74
Interorbital septum, 21, 65
Intestine, 39, 40, 41, 46
Iris, 81
Ischium, 33

JAWS, 8

Jugal bone, 22
Jugular veins, 44, 49

KIDNEY, 53

LACHRYMAL gland, 74

Large intestine, 40, 41, 46
Larynx, 57
Lateral ventricles, 85
Left aorta, 42, 45, 57; gastro-epiploic artery, 46
Lens, 81
Limbs, 9, 31, 34
Liver, 39, 40
Lobe, olfactory, 65, 82, 83; optic, 70, 84; prosencephalic, 65, 70, 82, 83
Lumbar plexus, 60; enlargement, 62
Lungs, 58
Lymph, 36; sinuses, 42

MALE reproductive organs, 50

Mandible, 26
Marginal plates, dermic, 12; epidermic, 10
Maxilla, 25
Meckel's cartilage, 26
Medulla oblongata, 66, 70, 82, 84
Membrane, nictitating, 7; tympanic, 8
Mesencephalon, 65, 70, 84
Mesenteric artery, 41
Mesentery, 40
Mesethmoid, 21
Metacarpus, 33
Metatarsus, 34
Midbrain, 65
Middle cervical ganglion, 86
Monro, foramen of, 85
Motor oculi (third cranial) nerve, 66, 70
Mouth, 7, 8
Müller's duct, 53
Muscle, attrahens pelvis, 36; external rectus, 73; inferior oblique, 74; inferior rectus, 74; internal rectus, 74; pectoral, 36; protractor of head, 60; pyramidalis, 74; retractor of head, 59; retractor of penis, 51; retrahens pelvis, 36; superior oblique, 64, 72; superior rectus, 66, 73

Myelon (spinal cord), 62
Myelon-cephalon (medulla oblongata), 66, 70, 82, 84

NARES, anterior, 7; posterior, 8, 19
Nasal chamber, 30, 64, 66
Naso-ethmoidal cartilage, 21, 81
Naso-prefrontal bone, 25
Nerves, abducens (fourth cranial), 67, 71, 76; auditory (eighth cranial), 67; caudal, 61; cervical spinal, 61; ciliary, 73; dorsal spinal, 61; facial (seventh cranial), 67, 71, 77; glosso-pharyngeal (ninth cranial), 71, 77, 78, 79; hypoglossal (twelfth cranial), 71, 77, 78; inferior maxillary, 75; motor oculi (third cranial), 66, 70; olfactory (first cranial), 70; ophthalmic, 65, 72, 73; optic (second cranial), 65, 70; orbito-nasal, 65, 72, 73; pathetic (fourth cranial), 65, 66, 70; pneumogastric (tenth cranial), 44, 67, 71; recurrent laryngeal, 44; sciatic, 48, 60; spinal accessory (eleventh cranial), 67, 71; suboccipital, 61; superior maxillary, 75; sympathetic, 44, 79; trigeminal (fifth cranial), 67, 71
Neural arches of dorsal vertebræ, 14; plates, 11; spines, 11, 14, 17
Neuro-central sutures, 15
Nictitating membrane, 7, 74
Nuchal plates, dermic, 11; epidermic, 10

OBTURATOR foramen, 34

Occipital condyle, 18
Oculo-motor (third cranial) nerve 66, 70
Odontoid process, 17
Œsophagus, 59
Olfactory lobe, 65, 70, 82, 83; nerve, (first cranial), 70
Opisthotic bone, 23, 29
Ophthalmic ganglion, 73; nerve, 65, 66, 67, 72, 73
Optic chiasma, 66, 82; lobe, 70, 84; nerve (second cranial), 66, 70, 82; thalamus, 84; tract, 66, 82

Orbit, 67
Orbito-nasal (ophthalmic) nerve, 65, 66, 67, 72, 73
Orbito-temporal fossa, 18
Organs of reproduction, female, 55; male, 50
Os articulare, 27; centrale, 33; febulare, 34; intermedium, 32, 34; magnum, 33; pubis, 33; radiale, 33; tibiale, 34; ulnare, 32
Ovary, 55
Oviduct, 55

PALATE bone, 25

Pancreas, 40, 41
Pancreatic duct, 41
Parietal bone, 24, 29
Pathetic (fourth cranial) nerve, 65, 66, 70
Pectoral arch, 31; muscles, 36
Pelvic arch, 33
Penis, 51
Pericardium, 35
Peritoneum, 36, 58
Phalanges of foot, 34; of hand, 33
Pineal gland, 70
Pisiform bone, 32
Pituitary body, 82
Plastron, 6, 16
Plates of carapace, 11, 12, 15; of epidermic skeleton, 10
Pleuro-peritoneal cavity, 35
Plexus, brachial, 61; choroid, 70; lumbar, 60; sciatic, 60
Pneumogastric (tenth cranial) nerve, 44, 67, 71, 77
Portal vein, 40
Post-auditory process, 18, 19
Posterior nares, 8, 19; vena cava, 54
Post-frontal bone, 22
Precoracoid, 31
Premaxilla, 25
Process, anterior clinoid, 24; coronoid, 26; odontoid, 17; post-auditory, 18, 19
Proötic bone, 23, 29
Prosencephalic lobes (cerebral hemispheres), 65, 70, 82, 83
Pterygoid bone, 24, 29
Pubic bone, 33; symphysis, 33
Pulmonary artery, 42, 48, 56; vein, 48

Pupil, 81
Pygal plates, dermic, 12; epidermic, 10
Pylorus, 39
Pyramidalis muscle of eye, 74; of pelvis, 36

QUADRATE bone, 19, 23
Quadrato-jugal bone, 22

RADIUS, 32

Rectum, 41
Recurrent laryngeal nerve, 44
Renal artery, 47; vein, 54; organs, 53
Reni-portal vein, 37, 38, 54
Reptilia, characters of, 2
Reproductive organs, female, 55; male, 50
Respiratory organs, 58
Retina, 81
Retractor muscles of head and neck, 59; of penis, 51
Retrahens pelvis, 36
Rhinencephalon (olfactory lobes), 65, 70, 82, 83
Ribs, 13
Right aorta, 42, 45, 57
Roots of spinal nerves, 63

SACRAL ribs, 14; vertebræ, 14

Sauropsida, characters of, 1
Scales, 10
Scaphoid bone, 33
Scapula, 31
Scapulo-precoracoid bone, 31
Sciatic artery, 48; nerve, 48, 60; plexus, 60
Sclerotic, 80
Second cranial (optic) nerve, 66, 70, 82
Sella turcica, 24
Semicircular canals, 69
Semilunar bone, 32
Sense organs, 64, 68, 80
Septum, narium, 8; interorbital, 21, 65
Seventh cranial (facial) nerve, 67, 71, 77

Sinus venosus, 39, 41
Sixth cranial (abducens) nerve, 67, 71, 76
Skeleton, 11
Skull, 18
Small intestine, 39, 40, 46
Spinal accessory (eleventh cranial) nerve, 67, 71; column, 12; cord, 62; ganglia, 63; nerves, 61; nerves, roots of, 63
Spines, neural, 11, 14, 17
Spleen, 41
Splenial bone, 27
Squamosal bone, 19, 22
Stellate ganglion, 86
Stomach, 39
Stylo-hyoid ligament, 57
Subclavian artery, 42; vein, 49
Suboccipital nerve, 61
Superior cervical artery, 43; cervical ganglion, 77; maxillary nerve, 75; mesenteric artery, 46; oblique muscle, 64, 72; pancreatico-duodenal artery, 46; rectus muscle, 66, 73; vena cava, 41, 49
Supra-occipital bone, 19, 23, 28
Surangular bone, 27
Sutures, neuro-central, 15; of carapace, 14
Sylvius, aqueduct of, 85
Sympathetic nerve, 44, 79, 86
Symphysis, ischial, 33; pubic, 33

TAIL, 9, 18

Tarsalia, 34
Tarsus, 34
Temporal fossa, 18, 20
Tenth cranial (pneumogastric) nerve, 44, 67, 71, 77
Testes, 52
Thalamencephalon, 66, 82
Thalami optici, 84
Third cranial (oculo-motor) nerve, 66, 70; ventricle of brain, 85
Thymus, 44
Thyroid artery, 43
Tibia, 34
Tongue, 8
Trabecula cranii, 21, 81
Trachea, 39, 57
Trapezium, 33
Trapezoid bone, 33

Trigeminal (fifth cranial) nerve, 67, 71
Tube, Eustachian, 8
Twelfth cranial (hypoglossal) nerve, 71, 77, 78
Tympanic membrane, 8, 68
Tympanum, 20, 68

ULNA, 32

Unciform bone, 33
Ureter, 55
Urethral groove, 51, 55
Urinary bladder, 38, 40, 52; organs, 52, 53, 54

VASA efferentia, 53

Vas deferens, 53
Vagus nerve (see pneumogastric)
Vein, anterior abdominal, 36, 40; caudal, 28; cava inferior, 54; cava superior, 41, 49; coronary, 49; facial, 49; femoral, 37; hepatic, 42; internal jugular, 44, 49; internal mammary, 50; of urinary bladder, 38; portal, 40; pulmonary, 48; renal, 54; reni-portal, 37, 38, 54; subclavian, 49
Vena cava inferior, 54; superior, 41, 49
Venous sinus, 39, 41
Ventricle of heart, 39, 41, 56
Ventricles, cerebral, 70, 84, 85
Vertebræ, caudal, 18; cervical, 16; dorsal, 12; sacral, 14
Vestibule, 69
Viscera, abdominal, 39
Vitreous humor, 81
Vomer, 25

WINDPIPE, 39, 57

XIPHIPLASTRON, 16

MACMILLAN & CO.'S PUBLICATIONS.

ELEMENTARY SCIENCE CLASS-BOOKS.

Astronomy.—ELEMENTARY LESSONS IN ASTRONOMY. With Coloured Diagram and numerous Illustrations. By J. NORMAN LOCKYER, F.R.S. . New Edition. 16mo. $1.25.

QUESTIONS ON LOCKYER'S ELEMENTARY LESSONS IN ASTRONOMY. By JOHN FORBES-ROBERTSON. 40 cents.

Physiology.—LESSONS IN ELEMENTARY PHYSIOLOGY. With numerous Illustrations. By T. H. HUXLEY, F.R.S. New Edition. 16mo. $1.10.

QUESTIONS ON HUXLEY'S PHYSIOLOGY FOR SCHOOLS. By T. ALCOCK. 40 cents.

Botany.—LESSONS IN ELEMENTARY BOTANY. By D. OLIVER, F.R.S. With numerous Illustrations. New Edition. 16mo. $1.10.

Chemistry.—LESSONS IN ELEMENTARY CHEMISTRY, INORGANIC AND ORGANIC. By HENRY E. ROSCOE, F.R.S., with numerous Illustrations. New Edition. 16mo. $1.10.

A SERIES OF CHEMICAL PROBLEMS, prepared with Special Reference to the above. By T. E. Thorpe, Ph.D. New Edition, with Key. 50 cents.

QUESTIONS ON CHEMISTRY. A Series of Problems and Exercises in Inorganic and Organic Chemistry. By Francis Jones, F.R.S.E. 18mo. 75 cents.

Practical Chemistry.—THE OWENS COLLEGE JUNIOR COURSE OF PRACTICAL CHEMISTRY. By FRANCIS JONES, F.R.S.E. New Edition. 70 cents.

Political Economy.—POLITICAL ECONOMY FOR BEGINNERS. By MILLICENT G. FAWCETT. New Edition. 18mo. 75 cents.

SCIENTIFIC CATALOGUE.

Elementary Science Class-books—*continued.*

Logic.—ELEMENTARY LESSONS IN LOGIC; Deductive and Inductive, with copious Questions and Examples, and a Vocabulary of Logical Terms. By W. STANLEY JEVONS, LL.D., M.A., F.R.S. New Edition. 18mo. 90 cents.

Physics.—LESSONS IN ELEMENTARY PHYSICS. By BALFOUR STEWART, F.R.S. With numerous Illustrations. New Edition. 16mo. $1.10.

QUESTIONS ON STEWART'S ELEMENTARY PHYSICS. By Prof. T. H. CORE. 50 cents.

Anatomy.—LESSONS IN ELEMENTARY ANATOMY. By ST. GEORGE MIVART, F.R.S. With upwards of 400 illustrations. 16mo. $1.50.

Steam.—AN ELEMENTARY TREATISE. By JOHN PERRY, Professor of Engineering. With wood-cuts. 18mo. $1.10.

Physical Geography.—ELEMENTARY LESSONS IN PHYSICAL GEOGRAPHY. By ARCHIBALD GEIKIE, F.R.S. With coloured Maps and numerous Illustrations. 16mo. $1.10.

QUESTIONS ON THE SAME. 40 cents.

Geography.—CLASS BOOK OF GEOGRAPHY. By C. B. CLARKE, M.A., F.G.S. New Edition, with eighteen coloured Maps. 16mo. 80 cents.

Natural Philosophy.—NATURAL PHILOSOPHY FOR BEGINNERS. By I. TODHUNTER, M.A., F.R.S. Part I. The Properties of Solid and Fluid Bodies. 18mo. 90 cents. Part II. Sound, Light, and Heat. 18mo. 90 cents.

Electricity and Magnetism. — ELEMENTARY LESSONS IN ELECTRICITY AND MAGNETISM. By Prof. SYLVANUS THOMPSON. With Illustrations. [*Immediately.*

Sound.—AN ELEMENTARY TREATISE. By W. H. STONE, M.B. With Illustrations. 18mo. 90 cents.

Economics.—THE ECONOMICS OF INDUSTRY. By A. MARSHALL, M.A., and MARY P. MARSHALL. 16mo. 70 cents.

MANUALS FOR STUDENTS.

Cossa.—GUIDE TO THE STUDY OF POLITICAL ECONOMY. By Dr. LUIGI COSSA. Translated from the Second Italian Edition. With a Preface by W. STANLEY JEVONS, F.R.S. 12mo. $1.25.

Fawcett.—A MANUAL OF POLITICAL ECONOMY. By HENRY FAWCETT, M.P. New Edition, revised and enlarged. 12mo. $2.65.

Flower.—AN INTRODUCTION TO THE OSTEOLOGY OF THE MAMMALIA. By Professor W. H. FLOWER, F.R.S., F.R.C.S. With numerous Illustrations. New Edition, enlarged. 12mo. $2.50.

Foster and Langley.—A COURSE OF ELEMENTARY PRACTICAL PHYSIOLOGY. By MICHAEL FOSTER, M.D., F.R.S., and J. N. LANGLEY, B.A. Fourth Edition. 12mo. $1.50.

Huxley.—PHYSIOGRAPHY. An Introduction to the Study of Nature. By Professor HUXLEY, F.R.S. With numerous Illustrations and coloured Plates. 12mo. $1.80.

Huxley and Martin.—A COURSE OF PRACTICAL INSTRUCTION IN ELEMENTARY BIOLOGY. By Professor HUXLEY, F.R.S., assisted by Prof. H. NEWELL MARTIN, M.B., D.Sc., Johns Hopkins University, Baltimore. 12mo. $1.50.

Jevons.—THE PRINCIPLES OF SCIENCE. A Treatise on Logic and Scientific Method. By Professor W. STANLEY JEVONS, LL.D., M.A., F.R.S. New and revised Edition. 12mo. $2.75.

STUDIES IN DEDUCTIVE LOGIC. By Professor W. STANLEY JEVONS. 12mo. $1.60.

Kiepert.—A MANUAL OF ANCIENT GEOGRAPHY. From the German of DR. H. KIEPERT. 12mo. $1.50.

Martin and Moale.—HANDBOOK OF VERTEBRATE DISSECTION. By H. NEWELL MARTIN, M.B., D.Sc., Johns Hopkins University, and WILLIAM A. MOALE, M.D. Part I. How to Dissect a Chelonian. With Illustrations. 12mo., 75c.

Manuals for Students—*continued.*

Parker.—A COURSE OF INSTRUCTION IN ZOOTOMY (Vertebrata). By T. JEFFREY PARKER, B. Sc. 12mo.
[*Immediately.*

Parker and Bettany.—THE MORPHOLOGY OF THE SKULL. By Professor PARKER and G. T. BETTANY. With Illustrations. 12mo. $2.60.

SCIENTIFIC TEXT BOOKS.

Balfour.—A TREATISE ON COMPARATIVE EMBRYOLOGY, with Illustrations. By F. M. BALFOUR, M.A., F.R.S., Fellow and Lecturer of Trinity College, Cambridge. In 2 vols. Vol. I., 8vo, $4.50. Vol. II., 8vo, $5.25.

Foster.—A TEXT-BOOK OF PHYSIOLOGY. By MICHAEL FOSTER, M.D., F.R.S. With Illustrations. Fifth Edition, revised. 12mo. Cloth, $3.00; sheep, $3.75.

Gamgee.—A TEXT-BOOK OF THE PHYSIOLOGICAL CHEMISTRY OF THE ANIMAL BODY. Including an account of the Chemical Changes occurring in Disease. By A. GAMGEE, M.D., F.R.S. In 2 vols., with Illustrations. Vol. I., 8vo, $4.50; sheep, $5.50.

Gegenbaur.—ELEMENTS OF COMPARATIVE ANATOMY. By Prof. CARL GEGENBAUR. A Translation by F. JEFFREY BELL, B.A. Revised with Preface by Prof. E. RAY LANKESTER, F.R.S. With numerous Illustrations. 8vo, $5.50; sheep, $6.50.

Geikie.—TEXT-BOOK OF GEOLOGY. By ARCHIBALD GEIKIE, F.R.S., Professor of Geology in the University of Edinburgh. With numerous Illustrations. 8vo. [*Immediately.*

Schorlemmer.—A MANUAL OF THE CHEMISTRY OF THE CARBON COMPOUNDS, OR ORGANIC CHEMISTRY. By C. SCHORLEMMER, F.R.S. 8vo. $3.75.

MACMILLAN AND CO., NEW YORK.

www.ingramcontent.com/pod-product-compliance
Lightning Source LLC
Chambersburg PA
CBHW020149170426
43199CB00010B/957